Structures and Functions of Retroviral RNAs

Nucleic Acids Set

coordinated by
Marie-Christine Maurel

Volume 1

Structures and Functions of Retroviral RNAs

The Multiple Facets of the Retroviral Genome

Philippe Fossé

iSTE

WILEY

First published 2022 in Great Britain and the United States by ISTE Ltd and John Wiley & Sons, Inc.

ISTE Ltd
27-37 St George's Road
London SW19 4EU
UK

www.iste.co.uk

John Wiley & Sons, Inc.
111 River Street
Hoboken, NJ 07030
USA

www.wiley.com

Library of Congress Control Number: 2022939512

British Library Cataloguing-in-Publication Data
A CIP record for this book is available from the British Library
ISBN 978-1-78630-826-9

Contents

Foreword

The sequencing of the human genome has turned our vision of biology upside down, revealing little by little the complexity of living organisms. We have realized that genes, DNA coding for proteins, do not control everything in the cell. Of the two meters of DNA in a cell, the genes represent only six centimeters. However, almost all of this DNA (80–90%) is transcribed into RNA. When they are not coding, these RNAs are regulators of the cell, and their role is strongly dependent on their structures. Unlike DNA, RNA is a dynamic molecule that can adopt different conformations, giving it several functions.

Sequencing has also revealed the strong presence of sequences of retroviral origin in the form of retroelements, retrotransposons or endogenous retroviruses, which are vestiges of our cohabitation with these viruses. Their roles remain mostly unknown: some are beneficial, for example by helping the placenta to form, and others trigger pathological processes, in particular cancerous ones. When these are able to leave the cell and infect other cells, they are called retroviruses. With the arrival of the Covid-19 pandemic caused by SARS-CoV-2, RNA viruses are now perceived as a real threat on a global scale. This pandemic is reminiscent of other pandemics, including acquired immunodeficiency syndrome (AIDS) caused by the human immunodeficiency virus (HIV) retrovirus. These pandemics highlight the strong need to understand the functioning mechanisms of these viruses and their ability to spread.

RNA, the probable ancestor of DNA and proteins, is a fascinating molecule because of its complexity, multifunctionality and remarkable adaptation to its host. This versatility of RNA is consistent with the multiple primary, secondary and tertiary structures it can adopt. Retroviral RNA is a perfect example with its numerous functions. Indeed, it plays several roles during infection: initially, it serves as a template for DNA synthesis through the reverse transcriptase, then as messenger RNA (mRNA) encoding structural and enzymatic proteins, and finally as the genome, the guardian of the virus genetic information (gRNA). It is able to hijack the cellular machinery of its host and transgress the rules of cellular RNA metabolism. For example, when used as mRNA, it can escape splicing and leave the nucleus despite the persistence of introns. It can also take two different export routes to exit the nucleus, presumably in different conformations. Recent studies have led to the discovery and understanding of the mechanisms of new nuclear export pathways, which are also used by some cellular RNAs. Retroviral RNAs have assisted in major discoveries and have shattered several biological dogmas.

This book reveals how RNA folds back on itself when it comes into contact with its partners encountered during the different stages of virus replication in the cell. We will learn how the structure of RNA evolves when it is copied into DNA by the viral reverse transcriptase after the virus enters the cell. One of the major questions of retrovirology is understanding how all the viral components (proteins and RNA), present in multiple copies, are found in a concerted manner, in number and time, at the periphery of the cell, in order to assemble and form new viruses, ready to disseminate outside the cell. To be infectious, a virus must contain not one but two copies of the same gRNA; we then speak of dimeric gRNA. How, among the multitude of cellular RNAs present in the cytoplasm of the cell, is the gRNA able to pair up and find its protein partners to form new viruses? Once again, the answer lies in the structure of this RNA.

Philippe Fossé, a distinguished director of research at the CNRS, focuses all his research on the study of the RNA structure of various retroviruses (avian, murine and human), with the aim of understanding the complex relationships between RNA structure and function. This is very meticulous work on a molecular scale, which requires extreme rigor and a certain insight acquired through experience. Indeed, computer programs for structure prediction, although increasingly sophisticated, are not by themselves sufficient to account for the versatility and folding dynamics of these RNAs.

Marylène MOUGEL
Institut de recherche en infectiologie
de Montpellier (IRIM)
May 2022

Preface

Retroviruses, which are single-stranded RNA viruses of positive polarity, have been identified in various vertebrate groups but not in invertebrates. Retroviruses originated with their aquatic vertebrate hosts at least 450 million years ago and have evolved through interactions with them. Retroviruses have contributed to vertebrate evolutionary processes. The most prominent example in host evolution is the formation of the placenta in the ancestors of placental mammals through several independent retroviral infections. Retroviruses are divided into two subfamilies (*Orthoretrovirinae* and *Spumaretrovirinae*). Spumaretroviruses infect a wide variety of mammals and are generally non-pathogenic. In contrast, retroviruses belonging to the *Orthoretrovirinae* subfamily are often responsible for pathologies in the vertebrates they infect. From the beginning of the 20th century until the beginning of the 1980s, in order to elucidate the mechanisms of carcinogenesis, numerous studies were focused on avian and murine retroviruses that induce leukemia and cancer in their hosts. These studies led to the discovery of oncogenes, contributing to the understanding of the regulation of eukaryotic gene expression and the characterization of the stages of the retrovirus replicative cycle. In addition, they have enabled fundamental scientific and technological advances in biology through the discovery of reverse transcriptase and the use of retroviral vectors in the analysis of gene expression. Since its discovery in 1983, HIV, the causative agent of AIDS, has been the main focus of retrovirology research.

Although the retrovirus genome is small (7–12 kb), in its RNA form, it performs multiple functions other than serving as a messenger for the synthesis of proteins necessary for the production of infectious viral particles. These functions, some of which vary according to the retroviral species, depend mainly on the structures adopted by retroviral RNA. In this book, which is based on the extensive scientific literature, I provide a non-exhaustive review of the knowledge acquired on the structure–function relationships of RNA in different retroviral species. In Chapter 1, I present general knowledge on retroviruses, as it is necessary to understand the molecular mechanisms regulated by RNA structures. Reverse transcription of retroviral RNA is a complex and essential process in the replicative cycle of retroviruses. The key steps of this process, which rely on interactions involving DNA and RNA structures, are presented in Chapter 2. Chapter 3 shows that, in some retroviruses, their genomic RNA forms secondary structures that serve as signals to regulate proviral DNA transcription, maturation, export and translation of retroviral RNA. Encapsidation is a process common to all retroviruses that allows the incorporation of two genomic RNA molecules into the viral particle. It requires interactions between several molecules of a retroviral protein and secondary structures of the genomic RNA, as described in Chapter 4.

For ease of reading, a glossary and a list of abbreviations and acronyms are included following the conclusion of this book. I hope that this book enables students and researchers to understand the multiple facets of retroviral RNA and contributes to developing their knowledge and critical thinking in the fields of research involving functional RNAs.

May 2022

1

General Information on Retroviruses

1.1. Common characteristics of retroviruses

Retroviruses, like all viruses, are parasites that lack the genetic information encoding the enzymes of intermediary metabolism and can therefore only replicate inside living cells. They infect vertebrates and can cause cancerous tumors, leukemia, neurological disorders and AIDS. Most retroviruses are exogenous and their transmission is achieved by contagion between distinct individuals. Others, known as endogenous, are integrated into the host genome and are transmitted hereditarily.

Although they are capable of infecting different animal host cells and causing different pathologies, all retroviruses have common structural and functional characteristics that allow them to be grouped in the family *Retroviridae* (see Table 1.1). In all retroviruses, the genetic information is carried by a single-stranded RNA.

The term retrovirus comes from the fact that their replication cycle imposes a passage from the RNA genome to a DNA form; *retro-* thus refers to the unusual direction from RNA to DNA. This passage is carried out by way of reverse transcriptase (RT), a retroviral enzyme that is an RNA- and DNA-dependent DNA polymerase (Baltimore 1970; Temin and Mizutani 1970).

The RNA genome of retroviruses, called genomic RNA (gRNA), consists of terminal non-coding regions necessary for viral replication and internal regions that code for viral enzymes and structural proteins (see Figure 1.1).

In the following, the generic term retrovirus only refers to retroviruses belonging to the subfamily *Orthoretrovirinae*.

Family	Subfamily	Type	Species
Retroviridae	*Orthoretrovirinae*	*Alpharetrovirus*	9 species
		Betaretrovirus	5 species
		Deltaretrovirus	4 species
		Espsilonretrovirus	3 species
		Gammetrovirus	18 species
		Lentivirus	10 species
	Spumaretrovirinae	*Spumavirus*	19 species

Table 1.1. *Classification of retroviruses*[1]

1.1.1. *Untranslated regions*

The untranslated terminal regions of gRNA are arranged in the same order in all retroviruses (see Figure 1.1), but depending on the species, they differ in terms of size, structure and some of their characteristics (Coffin 1992).

gRNA is derived from the same maturation process as cellular RNAs. A methylated guanosine m7G5' ppp5' is present at the 5' end of the 5' non-coding region; this cap is required for ribosome attachment and is important for the translation of viral RNAs that serve as messengers (Bolinger and Boris-Lawrie 2009). A poly(A) tail consisting of approximately 200 adenines is present at the 3' end of the 3' non-coding region (Vogt 1997).

1.1.1.1. *The 5' untranslated region (5'-UTR)*

This region is composed of four sequences (R, U5, PBS and L). The R (repeat) sequence, 16–247 nucleotides depending on the retrovirus, is present

1 Available from: https://talk.ictvonline.org/ictv-reports/ictv_9th_report/reverse-transcribing-dna-and-rna-viruses-2011/w/rt_viruses/161/retroviridae.

at both ends in all retroviruses (Klaver and Berkhout 1994). It plays an essential role in the replicative strategy of the retrovirus. The unique U5 sequence, 80–240 nucleotides depending on the retrovirus, is located downstream of the R sequence (Vogt 1997). It is the first region of the gRNA that is copied into DNA during reverse transcription. The primer binding site (PBS) sequence, consisting of 18 nucleotides, pairs with the 3' end of a cellular tRNA molecule that serves as a primer to start reverse transcription (Vogt 1997). The nature of the tRNA primer differs among retroviruses. The leader (L) sequence lies between the PBS and the *gag* gene and comprises at least part of the gRNA packaging signal in the virus particle.

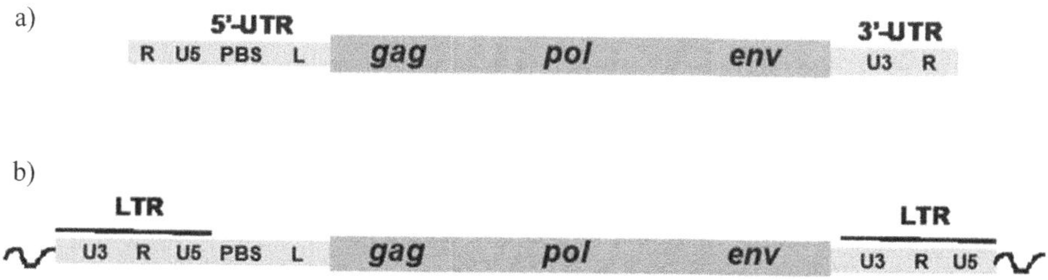

Figure 1.1. *Genetic organization of retroviruses. (a) Genomic RNA and (b) proviral DNA. The cellular DNA is represented by a wavy line. For a color version of this figure, see www.iste.co.uk/fosse/structures.zip*

1.1.1.2. *The 3' untranslated region (3'-UTR)*

This region is composed of three sequences (PPT, U3 and R). The polypurine tract (PPT) sequence, very rich in purines, 9–15 depending on the retrovirus, is located upstream of the U3 sequence. It serves as a primer for the synthesis of the (+) strand of the proviral DNA (Vogt 1997). The U3 sequence, a unique 3' sequence, is located between the PPT and the R sequence. U3 contains the promoter and regulator elements of viral RNA transcription. The R sequence is identical to that present in the 5'-UTR. It enables the first strand transfer, which is a crucial step in reverse transcription.

1.1.2. *Translated regions*

All retroviruses have the *gag*, *pol* and *env* genes (see Figure 1.1). The *gag* cistron and the *pol* cistron are often considered a single cistron, although there may be a phase shift between the two reading frames.

1.1.2.1. *The* gag *gene (group-specific antigen)*

This codes for a polyprotein precursor. This precursor is translated from gRNA which is an unspliced mRNA. During the maturation of the virus particle, the Gag precursor is cleaved by the viral protease (PR) to generate the structural proteins MA, CA and NC (see Figure 1.2).

1.1.2.2. *The* pol *(polymerase) gene*

The Gag–Pol polypeptide precursor is translated from gRNA when protein synthesis is not stopped at the stop codon of the *gag* gene. Depending on the retrovirus, two types of mechanisms can be used to continue translation after the stop codon: a reading frame-shift mechanism between the *gag* and *pol* genes (Jacks and Varmus 1985; Moore et al. 1987; Wilson et al. 1988; Nam et al. 1993) or the incorporation of an amino acid at the stop codon (Yoshinaka et al. 1985a; Yoshinaka et al. 1985b). The rate of synthesis of the Gag–Pol precursor is about 5% compared to that of the Gag precursor (Shehu-Xhilaga et al. 2001a). Proteolysis of Gag–Pol by PR is the origin of the structural proteins (MA, CA and NC) and viral enzymes: RT and integrase (IN) (Konvalinka et al. 2015).

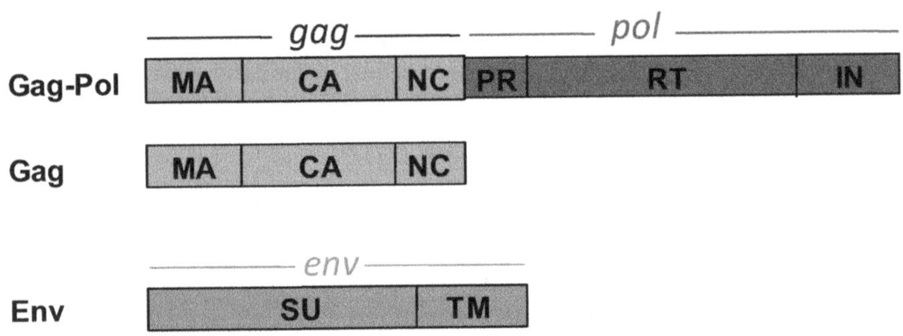

Figure 1.2. *Products of the* gag, pol *and* env *genes. For a color version of this figure, see www.iste.co.uk/fosse/structures.zip*

COMMENTARY ON FIGURE 1.2.– *Schematic representation of the three polypeptide precursors synthesized from the* gag, pol *and* env *genes. Although in all retroviruses the protease-encoding sequence (PR) is always between the* gag *and* pol *genes, its reading frame varies among retroviral species (Konvalinka* et al. *2015). For example, in HIV-1, PR is in the same reading frame as the* pol *gene, whereas it is in the reading frame of* gag *in alpharetroviruses and a separate reading frame in Mason–Pfizer monkey virus (MPMV).*

1.1.2.3. *The* env *(envelope) gene*

The *env* gene, unlike the *gag* and *pol* genes, is translated from spliced mRNA. The Env polypeptide precursor undergoes post-translational modifications, such as glycosylation, before being cleaved by a cellular protease into two subunits: the surface protein (SU) and the transmembrane protein (TM). The two subunits remain associated by non-covalent bonds in most retroviruses and form a trimer, with the portion exposed on the surface of the virus particle constituting the spike involved in host cell recognition (Steckbeck et al. 2014).

1.2. Architecture of the virion

In retroviruses, the infectious viral particle called the virion is spherical, 100–200 nm in diameter (Zhang et al. 2015). Retroviruses are enveloped viruses because the outer envelope of the virion is composed of a lipid bilayer. This is derived from the plasma membrane of the infected cell and is enriched in viral envelope protein (see Figure 1.3). The inner part of the virus forms a shell called the capsid, which is made up of self-assembling capsid proteins (CA). In all retroviruses belonging to the *Orthoretrovirinae* subfamily, the capsid contains the viral enzymes (PR, RT and IN) and the diploid genome of the virus consisting of two gRNA molecules in close association with the nucleocapsid (NC) proteins (Vogt 1997). The capsid also contains mRNA, tRNA and host cell proteins. The matrix consists mainly of matrix proteins (MA) and is located between the envelope and the capsid.

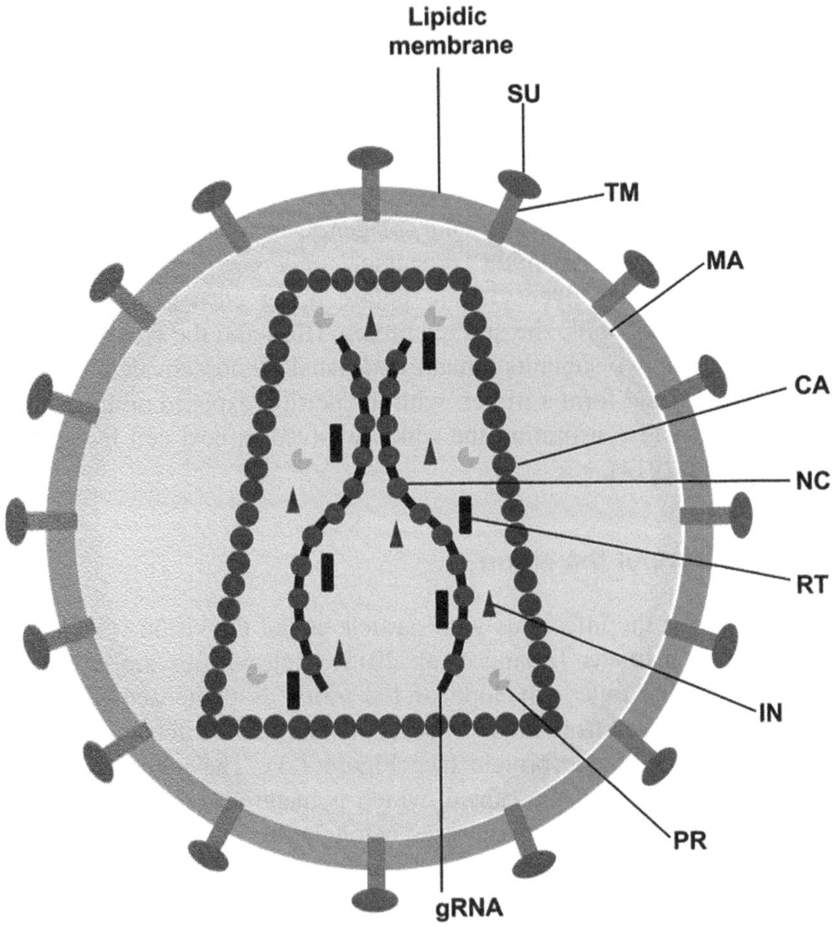

Figure 1.3. *Schematic representation of the virion. The structural organization of the virion is the same in all* Orthoretrovirinae. *The capsid containing viral enzymes and gRNA associated with NC molecules forms a closed shell with a morphology that varies among retroviral species (conical, tubular, nearly spherical or polyhedral) (Zhang et al. 2015). For a color version of this figure, see www.iste.co.uk/fosse/structures.zip*

1.3. Replication cycle of retroviruses

All retroviruses have an infection cycle with common stages. The replicative cycle of a retrovirus is divided into two main phases (D'Souza and Summers 2005): early and late (see Figure 1.4). The early phase includes the steps from binding the virion to the host cell receptor up to the integration of the double-stranded viral DNA into the host cell genome (proviral DNA). The late phase includes the following steps from transcription of the proviral DNA to the release of new virions into the extracellular medium.

1.3.1. *Early phase*

1.3.1.1. *Entry of the virus into the host cell*

The attachment of the virion to the target cell is mediated by a specific interaction between a receptor on the target cell and the envelope glycoprotein of the virus. There are several types of cellular receptors recognized by retroviruses; however, only one type of receptor is used by a retroviral species (Sommerfelt 1999). The envelope glycoprotein–receptor interaction triggers fusion of the cell and viral membranes and thus leads to the release of the capsid into the cytoplasm of the host cell (Lindemann et al. 2013). In the case of retroviruses belonging to the genus *Lentivirus*, such as HIV type 1 (HIV-1) and type 2 (HIV-2), an interaction of the envelope glycoprotein with a coreceptor is required for strong binding of the virion to the cell membrane and to allow fusion of the cell and viral membranes (Deng et al. 1996; Feng et al. 1996).

1.3.1.2. *Reverse transcription and decapsidation*

The reverse transcription of the gRNA, which is described in detail in Chapter 2, begins as soon as the capsid enters the cell. This complex process converts the single-stranded gRNA into double-stranded DNA, with a long repeating sequence (LTR) at both ends (see Figure 1.1). Capsid disassembly (decapsidation) has primarily been studied in HIV-1 and is partly dependent on reverse transcription (Dismuke and Aiken 2006; Arhel et al. 2007; Hulme et al. 2011; Cosnefroy et al. 2016; Mamede et al. 2017; Fernandez et al. 2019). There are three models of decapsidation:

– progressive decapsidation in the cytoplasm during transport to the nuclear membrane (Mamede et al. 2017);

– decapsidation at a nuclear pore (Fernandez et al. 2019);

– decapsidation after entering the nucleus of an intact or nearly intact capsid (Burdick et al. 2020).

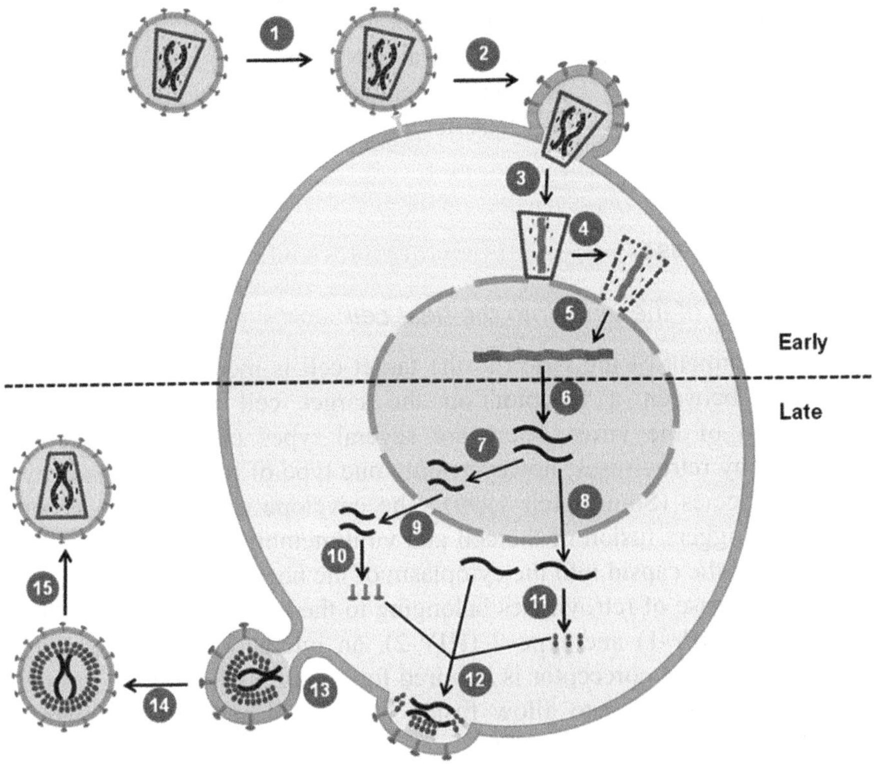

Figure 1.4. *Replicative cycle of a retrovirus. For a color version of this figure, see www.iste.co.uk/fosse/structures.zip*

COMMENTARY ON FIGURE 1.4. – *(1) Attachment of the virion to the target cell. (2) Fusion of cell and viral membranes. (3) Reverse transcription and transport of the capsid to the nuclear membrane. (4) Decapsidation at a nuclear pore. (5) Entry of double-stranded viral DNA (orange band) into the nucleus and integration into cellular DNA (gray band). (6) Transcription of the proviral DNA. (7) Splicing of a fraction of the gRNA molecules. (8) Nuclear export of a fraction of the gRNA molecules. (9) Nuclear export of the viral mRNA. (10) Translation of viral mRNAs. (11) Translation of gRNA molecules. (12) Assembly of a viral particle involving two gRNA molecules, SU and TM proteins and Gag and Gag–Pol precursors. (13) Budding of the viral particle. (14) Release of an immature virus particle into the extracellular environment. (15) Maturation of the viral particle.*

1.3.1.3. *Integration of the viral genome into the DNA of the host cell*

The integration of double-stranded viral DNA into the host cell genome, leading to proviral DNA (see Figure 1.1), is an essential step in the replicative cycle of retroviruses. It allows the virus, on the one hand, to express its genes using the cellular machinery of transcription and splicing, and on the other hand, to maintain its genome in the cells during cell divisions. This integration process requires IN, an enzyme encoded by the retrovirus *pol* gene, and cellular enzymes (Ciuffi 2016).

1.3.2. *Late phase*

1.3.2.1. *Synthesis, maturation and export of viral RNA*

Once integrated into the cellular genome, the proviral DNA is transcribed into RNA by the cell's transcription machinery. An m7G5'ppp5' cap and a poly(A) tail are added to the 5' and 3' ends of the newly synthesized RNA, respectively, corresponding to the retrovirus gRNA (Vogt 1997; Bolinger and Boris-Lawrie 2009). A fraction of the gRNA population is not spliced and is exported to the cytoplasm. From a fraction of the gRNA population, the cellular splicing machinery generates viral mRNAs that are exported to the cytoplasm. In all retroviruses, the gRNA contains at least one intron bounded by a splice donor site and a splice acceptor site to generate the *env* mRNA. The number of mRNA isoforms generated by alternative splicing of the gRNA varies among retroviruses. In HIV-1, alternative splicing can generate up to 100 multi-spliced RNAs (Ocwieja et al. 2012). The role of RNA structures in viral RNA synthesis, maturation and export is discussed in detail in Chapter 3.

1.3.2.2. *Translation of viral RNA*

After being exported to the cytoplasm, a fraction of the gRNA population is encapsidated into a forming virus particle and/or translated into Gag and Gag–Pol polypeptide precursors, which participate in virion formation. The SU and TM subunits of the envelope glycoprotein are synthesized from the *env* mRNA and integrated into the virus particle. Some of the proteins synthesized from the other viral mRNAs are also incorporated into the virion (e.g. Vpr in the case of lentiviruses) (Fabryova and Strebel 2019). The RNA structures and mechanisms that regulate protein synthesis are described in Chapter 3.

1.3.2.3. *Formation of the virion*

Two gRNA molecules are specifically encapsidated in the budding virus particle (see Figure 1.4). The encapsidation of the viral RNA, which requires the Gag precursor, is an obligatory step in the infectious cycle (D'Souza and Summers 2005). The RNA structures and mechanisms involved in this step are described in Chapter 4. The gRNA and Gag polypeptide play an important role in the assembly of new virus particles (D'Souza and Summers 2005; Olson et al. 2015). With the exception of betaretroviruses, assembly and budding of a new viral particle occur simultaneously at the plasma membrane (D'Souza and Summers 2005; Prchal et al. 2013). Gag and Gag–Pol polypeptides bind to the plasma membrane through their N-terminus. The presence at the plasma membrane of the Gag, Gag–Pol precursors, SU and TM proteins and two copies of the gRNA triggers the assembly and budding of a new virus particle. The virus particle is not immediately infectious after being released into the extracellular medium. It undergoes a maturation process that allows it to become infectious (virion). The maturation process is initiated by the viral PR which cleaves the Gag and Gag–Pol precursors. It begins during or just after the release of the virus particle (Pornillos and Ganser-Pornillos 2019).

2

Effects of the Structure of Retroviral RNA on Reverse Transcription

2.1. Reverse transcription of genomic RNA

Thanks to its DNA/RNA-dependent DNA polymerase activity, RT is the main actor in the reverse transcription of genomic RNA. For this reason, it is one of the main targets of pharmacological agents used against HIV-1. RT has no proofreading activity. Therefore, there is a particularly high rate of mutation during reverse transcription. In conjunction with the recombination events occurring during reverse transcription, mutations give retroviruses a high degree of genetic variability that leads them to evade host immune defenses and develop resistance to antiretroviral agents. The NC is the other viral protein that plays an important role in reverse transcription (Levin et al. 2010). This protein, which is not an enzyme, is rich in basic amino acids, and has one or two zinc fingers depending on the retroviral species (see Figure 2.1). It is a chaperone protein of nucleic acids, that is, it allows a nucleic acid to form the secondary structure that is thermodynamically most stable, which generally corresponds to the conformation with the maximum number of base pairs (Thomas and Gorelick 2008; Darlix et al. 2011). The chaperone activity of the NC is the result of three properties (Levin et al. 2010):

– nucleic acid aggregation, which is mainly due to basic amino acids;

– destabilization of short nucleic acid duplexes by zinc fingers;

– short residence time on the nucleic acid.

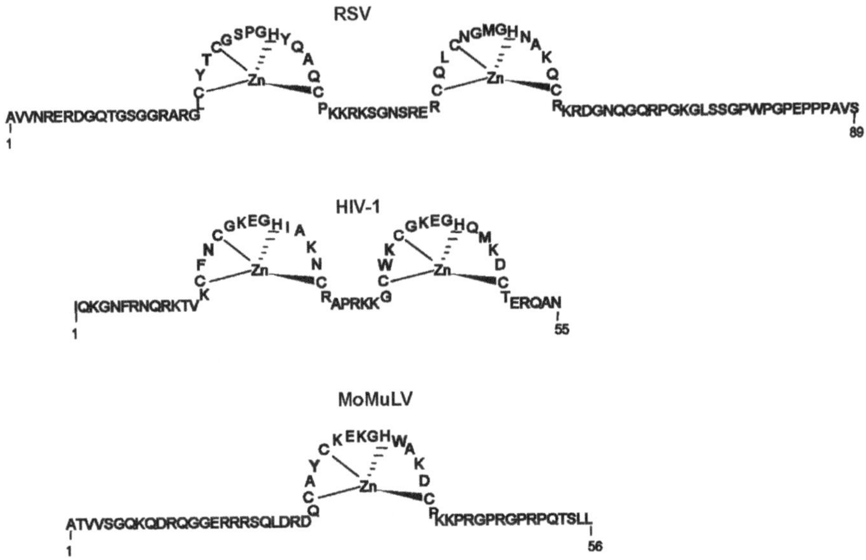

Figure 2.1. *Examples of NC sequences. NCs are small proteins (less than 100 amino acids) that have one (for example MoMuLV) or two (for example RSV and HIV-1) zinc fingers (source: Genbank: MoMuLV (J02255), HIV-1 (AF324493), RSV (J02342))*

Although in a few virions, reverse transcription of gRNA is initiated before entry into the cell, and it primarily begins when the infectious particle enters the cell (Trono 1992; Zhang et al. 1993; Hooker and Harrich 2003). The reverse transcription of the single-stranded gRNA is characterized by a sequence of steps leading to the synthesis of double-stranded DNA with an LTR at each end (Gilboa et al. 1979) (see Figure 2.2). Reverse transcription of a single gRNA molecule is theoretically sufficient to produce a double-stranded DNA molecule. However, studies suggest that the two gRNA molecules present as a dimer in the capsid are used to generate a single molecule of a double-stranded DNA molecule (Panganiban and Fiore 1988; Yu et al. 1998). The RNA structures that are involved in certain steps of reverse transcription will be described below.

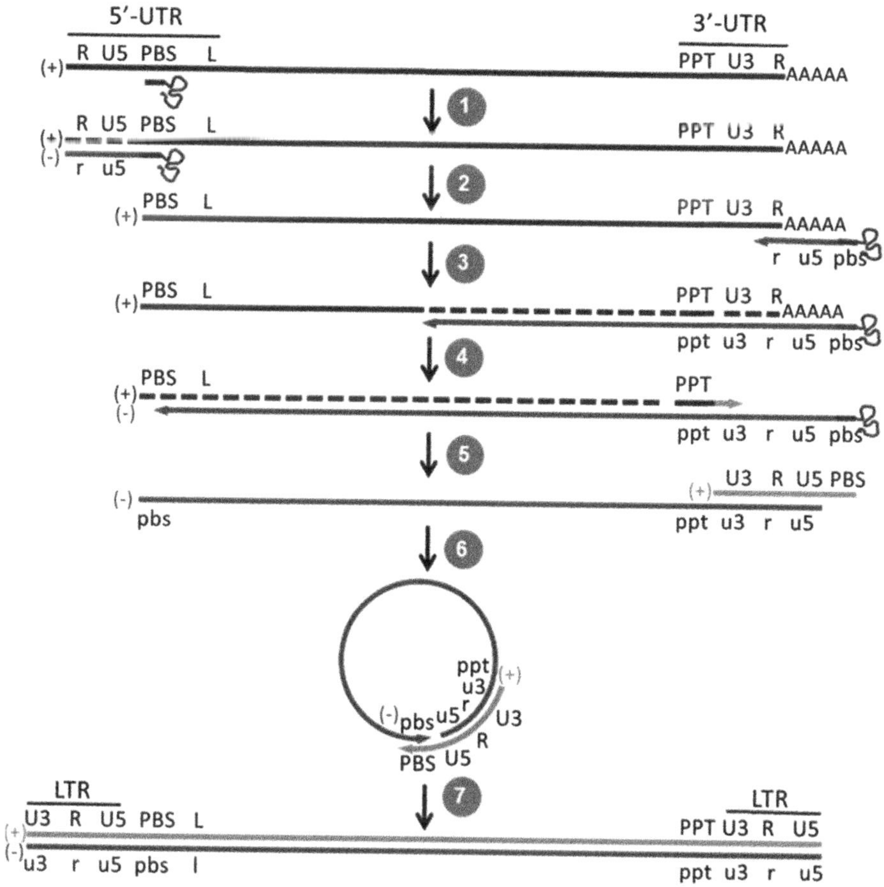

Figure 2.2. *Reverse transcription of the retroviral genome. For a color version of this figure, see www.iste.co.uk/fosse/structures.zip*

COMMENTARY ON FIGURE 2.2 – *In this diagram, only one of the two strands of the gRNA dimer is shown to facilitate the understanding of the different steps. (1) RT synthesis of negative polarity ssDNA from the tRNA primer bound to the PBS sequence. (2) RT RNase H degradation of the copied R-U5 region and first strand transfer at the R–r sequences. (3) Synthesis of the negative polarity DNA strand and degradation of the copied RNA strand by RT RNase H activity. (4) Further synthesis of the negative polarity DNA strand and initiation of the positive polarity DNA strand synthesis from the PPT RNA primer. (5) Complete synthesis of the negative polarity DNA. (6) Second strand transfer at the PBS–pbs complementary sequences. (7) The (+) and (-) DNA strands are extended by RT to produce double-stranded retroviral DNA with an LTR at each end. The retroviral genomic RNA is shown as a blue line. The tRNA primer is shown with a purple line. The DNA strands (-) and (+) are, respectively, shown with brown and orange lines. The direction of RT polymerization is shown with an arrowhead. The degradation of RNA by the RNase H activity of RT is shown with a dashed line. Sequence names are written in capital letters for the (+) strand and in lower case for the (-) strand.*

2.2. RNA structures involved in the initiation of reverse transcription

2.2.1. *A cellular tRNA as RT primer*

Like all DNA polymerases, RT needs a primer to start DNA polymerization. In retroviruses, the primer is a cellular tRNA (Marquet et al. 1995) which is $tRNA^{Lys3}$ in lentiviruses (e.g. HIV-1), $tRNA^{Trp}$ in alpharetroviruses (e.g. Rous sarcoma virus (RSV)) and $tRNA^{Pro}$ in murine retroviruses (e.g. Moloney murine leukemia virus (MoMuLV)) (see Figure 2.3).

The tRNA primer is incorporated into the virion as it is assembled. Incorporation of tRNA can be highly selective (e.g. in RSV and HIV-1) or minimally selective (e.g. in MoMuLV) (Mak and Kleiman 1997). Selective incorporation has been extensively studied in HIV-1; this involves the polypeptide precursors Gag and Gag–Pol, and Lysyl–tRNA synthetases of the host cell (Kaminska et al. 2007; Kleiman et al. 2010).

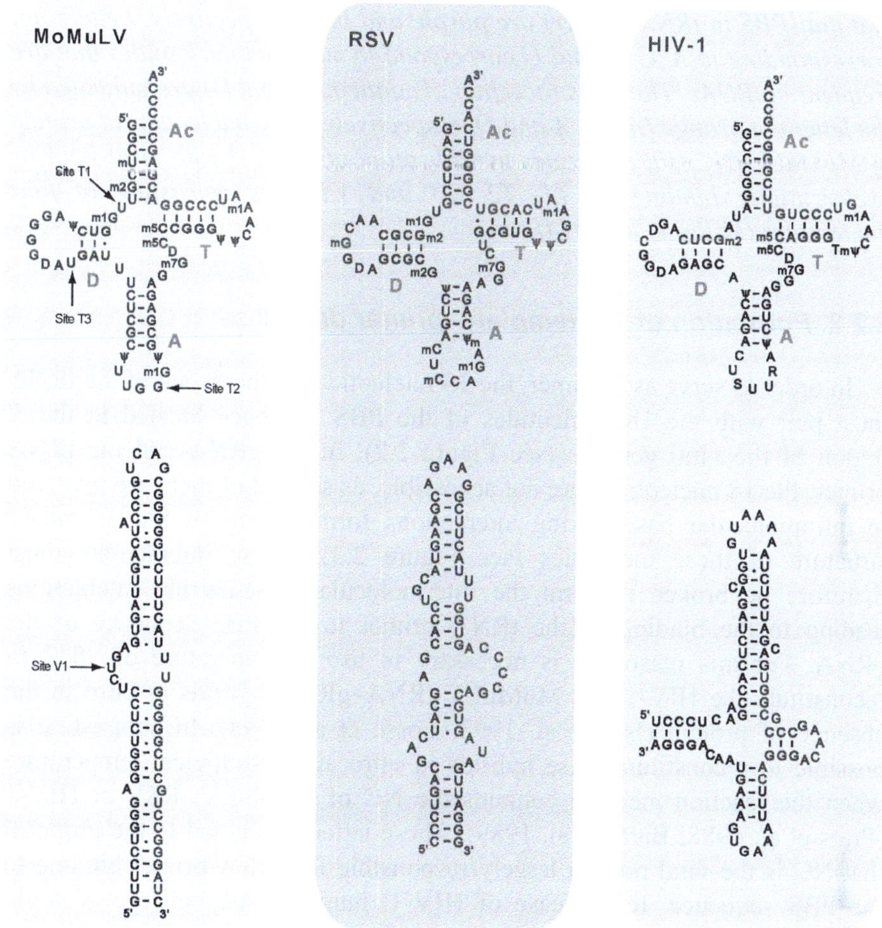

Figure 2.3. *Examples of secondary structures of tRNAs and gRNAs involved in the PBS–primer interaction. For a color version of this figure, see www.iste.co.uk/fosse/structures.zip*

COMMENTARY ON FIGURE 2.3.– *For each of the three retroviruses taken as examples, the secondary structure of the tRNA primer it uses is shown in the upper part of the figure, while the lower part shows the secondary structure of the portion of the gRNA that contains the PBS sequence (Keith and Heyman 1990; Hackett et al. 1991; Baudin et al. 1993; Mougel et al. 1993). Complementary sequences consisting of 18 nucleotides (PBS in viral RNA* *and anti-PBS in tRNA primer) are purple and blue, respectively. Letters not corresponding to A, C, G and U correspond to modified nucleotides that are frequent in tRNAs. The arms (acceptor, T, anticodon and D) are indicated by the letters in orange (Ac, T, A and D, respectively). The nucleocapsid protein of MoMuLV (NCp10) facilitates hybridization of tRNA^{Pro} with viral RNA by preferentially binding to T1, T2, T3 and V1 sites consisting of four nucleotides (Miller et al. 2014).*

2.2.2. *Formation of the template–primer duplex*

In order to serve as a primer, the 18 nucleotides at the 3' end of the tRNA must pair with the 18 nucleotides of the PBS sequence located in the 5' region of the viral genome (see Figure 2.2). In the gRNA and the tRNA primer, the 18 nucleotides are not accessible, as several of them are involved in intramolecular base-pairing interactions forming part of the secondary structure of these molecules (see Figure 2.3). These interactions must therefore be broken to form the intermolecular base-pairing interactions leading to the binding of the tRNA primer to the PBS sequence of the gRNA. For this reason, it is necessary to use a temperature of 70°C to reconstitute the HIV-1 and MoMuLV tRNA–gRNA hybrids in vitro in the absence of protein (Isel et al. 1993; Fossé et al. 1998). In contrast, it is possible to reconstitute these hybrids in vitro at physiological temperature when the reaction medium contains the NC of MoMuLV, RSV or HIV-1 (Prats et al. 1988; Barat et al. 1989). These latter results led to the proposal that NC is the viral protein largely responsible for tRNA primer binding to the PBS sequence. In the case of HIV-1, human RNA helicase A could facilitate hybridization by inducing a conformational change in the 5'-UTR region of the gRNA (Xing et al. 2011, 2012).

Nuclear magnetic resonance (NMR) analysis of the complexes that the MoMuLV NC forms with tRNA^{Pro} and the PBS region has determined the mechanism of formation of the tRNA^{Pro}–PBS region duplex in MoMuLV (Miller et al. 2014). Upon binding to the V1 site of the viral RNA and the T1 site of the tRNA^{Pro} (see Figure 2.3), the NC releases three nucleotides from the PBS sequence and also three nucleotides from the anti-PBS sequence of the tRNA^{Pro} to initiate the base-pairing interactions of the primer–template duplex. In addition, NC binding to the T2 and T3 sites destabilizes the three-dimensional structure of the $tRNA^{Pro}$, and thus increases the accessibility of the anti-PBS sequence. The hybridization of $tRNA^{Lys3}$ to the HIV-1 PBS sequence has also been studied by NMR (Tisné et al. 2004; Barraud et al. 2007). These studies indicate that in the anti-PBS sequence of $tRNA^{Lys3}$, a few nucleotides in the T arm and the 3' CCA end initiate the formation of the primer–template duplex by base-pairing. This initiation process does not require the presence of the NC, but it accelerates it. In contrast, the formation of the long PBS-anti-PBS duplex (18 base pairs) requires the action of the NC which destabilizes the three-dimensional structure of the $tRNA^{Lys3}$ by binding to the D-arm loop.

The distinct roles of the Gag precursor and NC on tRNA hybridization to the PBS sequence have primarily been studied in HIV-1 (Feng et al. 1999; Guo et al. 2009; Seif et al. 2015; Jin and Musier-Forsyth 2019). The secondary structures of the hybrid within mature and immature viral particles can be studied using the SHAPE method (Wilkinson et al. 2008; Seif et al. 2015). From these studies, a two-step hybridization model was proposed (see Figure 2.4):

– the formation of an unstable duplex, which is facilitated by the NC domain of the Gag precursor prior to its proteolysis during the maturation of the virus particle;

– the NC in the mature virus particle facilitates the conversion of the unstable duplex to a stable duplex.

The secondary structure model of the stable duplex shows that the $tRNA^{Lys3}$–gRNA interaction is not limited to the PBS sequence and also involves two additional interactions consisting of short sequences (four and six nucleotides). Interestingly, the hybrid reconstituted in vitro at 70°C in the absence of protein can mirror that present in the virion. Indeed, under these conditions, all three interactions between $tRNA^{Lys3}$ and gRNA have also been observed (Isel et al. 1995). It should be noted, however, that the two additional interactions would only occur in some HIV-1 isolates (Sleiman et al. 2012). In the case of RSV, the $tRNA^{Trp}$–gRNA interaction would not be limited to the PBS sequence but would also involve another sequence consisting of seven nucleotides (Aiyar et al. 1992; Morris et al. 2002). In contrast, the $tRNA^{Pro}$–gRNA interaction is limited to the PBS sequence in MoMuLV (see Figure 2.5) (Fossé et al. 1998).

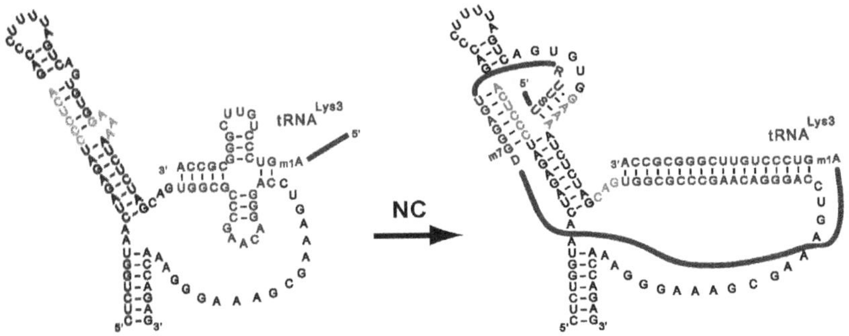

Figure 2.4. *Secondary structures of the HIV-1 tRNA^{Lys3}–gRNA hybrid. For a color version of this figure, see www.iste.co.uk/fosse/structures.zip*

COMMENTARY ON FIGURE 2.4. – *In HIV-1, the Gag precursor facilitates the formation of an unstable hybrid between the tRNA^{Lys3} and the PBS sequence (secondary structure on the left). The NC generated by Gag maturation converts the unstable hybrid to a stable hybrid (Seif et al. 2015). The PBS sequence is only partially paired with the 3′ end of the tRNA^{Lys3} in the unstable hybrid. The tRNA^{Lys3} is in purple; the nucleotides shown are mainly those that pair with the gRNA, while the others are replaced by the thick line. The PBS sequence is in blue, while the other two sequences of the gRNA that interact with the tRNA^{Lys3} are in green. The first three nucleotides to be copied by RT are in orange.*

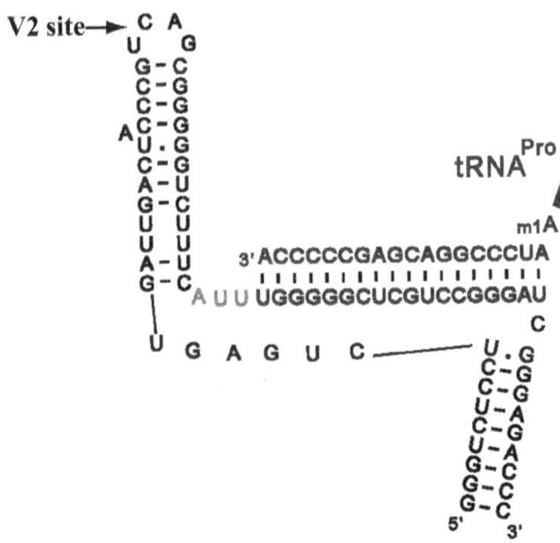

Figure 2.5. *Secondary structure of the MoMuLV–gRNA^{Pro} hybrid. For a color version of this figure, see www.iste.co.uk/fosse/structures.zip*

COMMENTARY ON FIGURE 2.5. – *The PBS sequence of the MoMuLV gRNA is fully paired with the 3' end of the tRNA^{Pro}. The tRNA^{Pro} is in purple; the nucleotides shown are mainly those that pair with the gRNA, while the others are replaced by the thick line. The PBS sequence is in blue. The first three nucleotides to be copied by RT are in orange. By binding to the V2 site, NC destabilizes the apical stem and thus facilitates extension of the tRNA^{Pro} primer by RT.*

2.2.3. *Role of the structure of the template–primer duplex upon initiation*

The determination of the secondary structure of the HIV-1 and MoMuLV duplexes (Isel et al. 1995; Fossé et al. 1998), using chemical and enzymatic probes, has shown that the first three nucleotides to be copied by RT are unpaired and adjacent to a stem-loop (see Figures 2.4 and 2.5). Binding of the MoMuLV NC to the apical loop corresponding to the V2 site (see Figure 2.5) facilitates extension of the tRNA^{Pro} primer by RT by destabilizing the apical stem of the viral RNA (Miller et al. 2014).

The reverse transcription initiation step corresponding to a short primer extension has been studied in detail in HIV-1 (Isel et al. 2010; Sleiman et al. 2012). This step is characterized by RT pauses during primer extension, rapid dissociation of RT from the primer–template duplex and a DNA polymerization rate that is slow. RT pausing leads to the accumulation of two intermediates corresponding to the addition of three and five nucleotides to the 3' end of the tRNA^{Lys3} primer, respectively. The elongation step of reverse transcription, which is fast and processive, starts as soon as the sixth nucleotide is added to the elongated primer.

The interaction of RT with the primer–template duplex during initiation can be characterized at the molecular level by a single-molecule study based on fluorescence resonance energy transfer (FRET) (Liu et al. 2010). This study shows that RT can bind to the primer–template duplex in two opposite orientations: one competent for DNA synthesis because the polymerase site of the enzyme is at the 3' end of the primer and one inactive for DNA synthesis because the polymerase site is opposite to the 3' end of the primer (see Figure 2.6). In the absence of deoxyribonucleotides, both binding modes are observed.

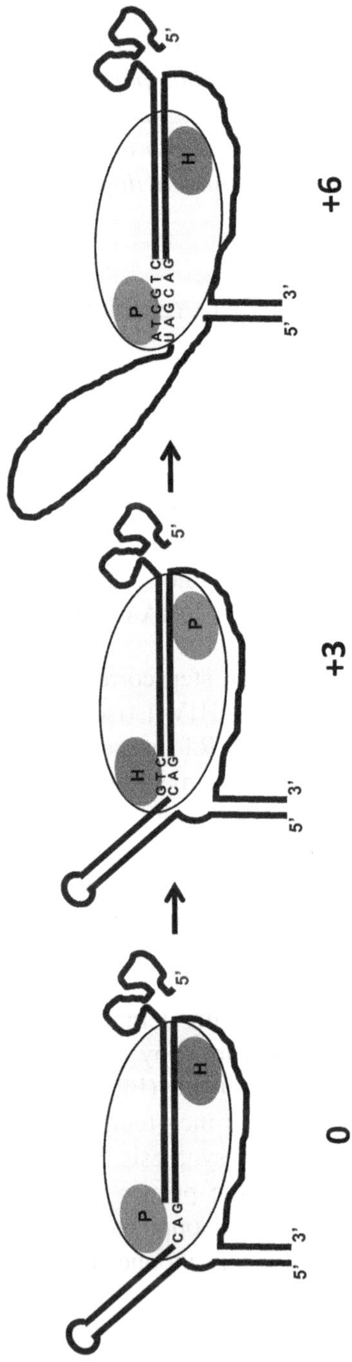

Figure 2.6. *HIV-1 RT orientations on the primer–template duplex. For a color version of this figure, see www.iste.co.uk/fosse/structures.zip*

COMMENTARY ON FIGURE 2.6. – *On the primer–template duplex consisting of the gRNA and tRNA^{Lys3} PBS sequence, the HIV-1 RT can adopt two opposite orientations in the absence of deoxyribonucleotides (state 0). In the diagram, only one orientation is shown in state 0, where the RT polymerase site (P) is near the 3' end of the tRNA^{Lys3} (purple molecule). After adding three (state +3) or five (state +5 not shown here) deoxyribonucleotides, the RT adopts an orientation that is not competent for DNA polymerization, because it is its RNase H (H) site that is close to the 3' end of the primer extended by three or five nucleotides. By adding the sixth nucleotide (+6 state), the RT adopts the competent orientation for DNA synthesis and thus begins to start the elongation step of reverse transcription, which is coupled with the destabilization of the stem-loop template (blue molecule).*

In the presence of deoxyribonucleotides, the orientation competent for DNA synthesis is promoted until three nucleotides are incorporated at the end of the primer. This three-nucleotide extension increases the likelihood of RT pausing because it allows the stem-loop located upstream of the primer to impose the inactive orientation of the enzyme for DNA polymerization. This orientation is prevalent until the incorporation of the fifth nucleotide. Incorporation of the sixth nucleotide destabilizes the stem-loop, and thus allows the RT to adopt the orientation competent for polymerization (see Figure 2.6). The destabilization of the stem-loop is thought to be facilitated by the NC (Wilkinson et al. 2008; Liu et al. 2010).

2.3. RNA structures involved in the first strand transfer

2.3.1. *Actors involved in the first strand transfer*

The first strand transfer, which is an essential step in reverse transcription, occurs at the R repeat sequence that is present at both ends of the gRNA (see Figure 2.2, step 2). The length of the R sequence varies by retroviral species (e.g. 16 nucleotides in mouse mammary tumor virus (MMTV) and 247 nucleotides with human T-cell leukemia virus type 2 (HTLV-2)) (René et al. 2018). The transfer primarily corresponds to a base-pairing interaction between the R region located at the 3' end of the gRNA and the r region of the newly synthesized DNA strand, called strong-stop DNA (ssDNA). The ssDNA is synthesized by the RT which copies the RU5 region of the gRNA. The first strand transfer thus requires at least three players: gRNA, RT and ssDNA. RNase H activity of RT, leading to degradation of the RU5 RNA region that has been copied, is required for the transfer reaction because it releases the ssDNA (Tanese et al. 1991; Peliska and Benkovic 1992; Blain and Goff 1995; Chen et al. 2003). Simplified experimental systems using RT, two short RNAs synthesized in vitro and a DNA primer have been able to efficiently reproduce the first strand transfer in vitro when the R region is truncated (Peliska and Benkovic 1992; Werner et al. 2001). In contrast, the experimental system must contain the NC to obtain in vitro efficient transfer when both RNAs are relatively long and contain an entire R sequence (Darlix et al. 1993; Allain et al. 1994; Guo *et al.* 1997). Under these conditions, NC increases the transfer by five- to tenfold (Guo et al. 1997; Chen et al. 2003). In addition to RT, NC thus appears to be another viral protein playing an important role in the first strand transfer.

2.3.2. *Molecular basis of R–r pairing*

The molecular mechanisms governing R–r pairing between the 3' end of gRNA and the 3' end of ssDNA have been primarily studied in HIV-1. In vitro analysis of ssDNA with a chemical probe and two enzyme probes suggests that NC binding to four preferential sites triggers unfolding of the ssDNA three-dimensional structure and thus facilitates the interaction of the r-sequence with the gRNA R-sequence (Chen et al. 2016b). An in vitro study (Kanevsky et al. 2005) and an ex vivo study using SHAPE (Watts et al. 2009) have shown that the 3' R sequence of the gRNA folds to form two stem-loops called TAR and poly(A). It has also been shown in vitro (Chen et al. 2016b) that the r-sequence of the ssDNA forms the cTAR and cpoly(A) stem-loops that are complementary to the TAR and poly(A) stem-loops, respectively (see Figure 2.7). The unfolding of the secondary structures constituted by these stem-loops is also necessary to allow r–R pairing corresponding to a stable duplex of 95 base pairs.

In vitro and ex vivo studies suggest that TAR/cTAR stem-loops play a more important role in r–R pairing than poly(A)/cpoly(A) stem-loops (Ohi and Clever 2000; Berkhout et al. 2001; Moumen et al. 2001). The interaction of NC with the TAR RNA stem-loop can be characterized in vitro at the molecular level by a single-molecule study using optical tweezers (McCauley et al. 2015) and by an NMR study (Belfetmi et al. 2016). The single-molecule study showed that NC can destabilize the TAR stem-loop by targeting four paired guanines that are adjacent to low-stability regions (bulges and G-U pair) (see Figure 2.8a).

NMR has shown that among these guanines, the one located in the apical stem plays the most important role, as it corresponds to the site where the NC initiates the destabilization of the TAR stem-loop.

Biochemical and biophysical studies performed in vitro with the cTAR stem-loop in the absence of NC (Bernacchi et al. 2002; Azoulay et al. 2003; Vo et al. 2009; Kanevsky et al. 2011) and ssDNA (Chen et al. 2016b) revealed a dynamic structure for this stem-loop, which results from the equilibrium between two conformations (see Figure 2.8b).

The same studies have shown that the NC shifts the equilibrium to the more open conformation (see Figure 2.8b).

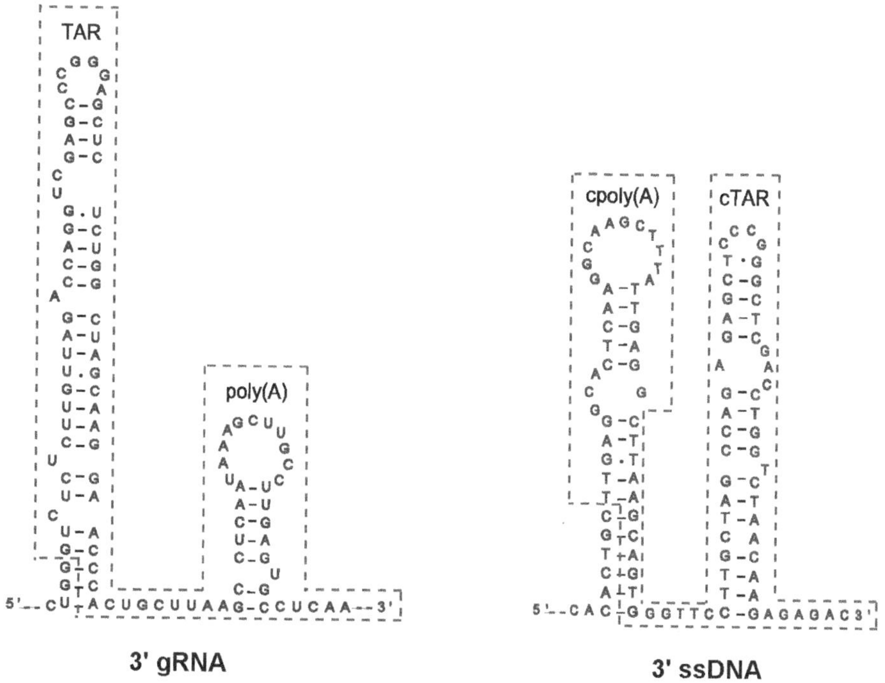

Figure 2.7. *Secondary structures of the 3′ ends of gRNA and ssDNA. Secondary structures were determined using chemical and enzymatic probes with the HIV-1 MAL isolate (Kanevsky et al. 2005; Chen et al. 2016b). The complementary r and R sequences are framed by the dashed lines. For a color version of this figure, see www.iste.co.uk/fosse/structures.zip*

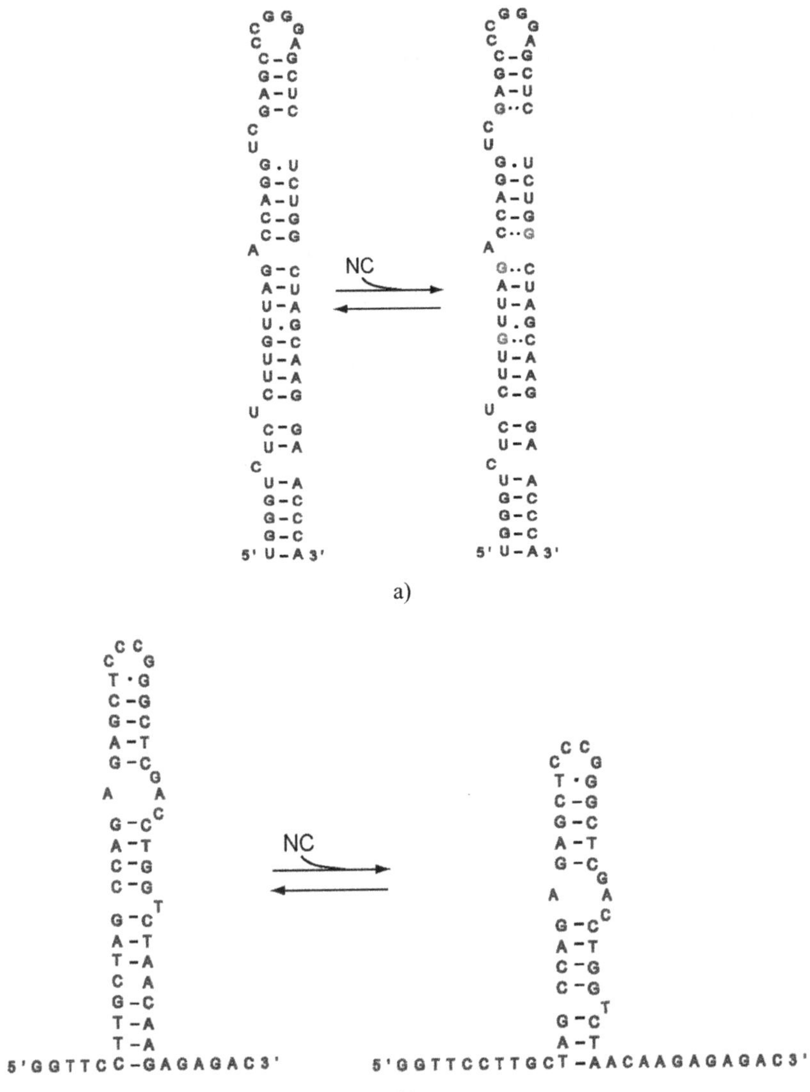

Figure 2.8. *a) Effect of NC on the TAR stem-loop; b) effect of NC on the cTAR stem-loop. For a color version of this figure, see www.iste.co.uk/fosse/structures.zip*

COMMENTARY ON FIGURE 2.8. – *NC destabilizes the TAR RNA stem-loop shown in panel A by preferentially interacting with the four guanines colored in orange and red. The destabilized G–C base pairs are outlined by a dashed line. Destabilization is initiated at the red-colored guanine. The NC destabilizes the cTAR DNA stem-loop presented in panel B by destabilizing four base pairs, thus enabling the opening of the bottom of the stem.*

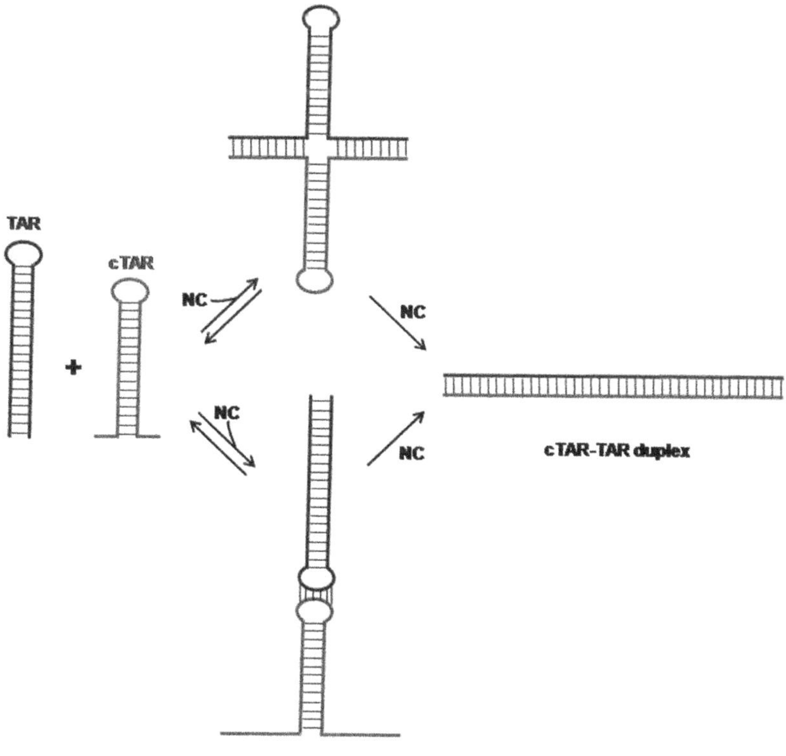

Figure 2.9. *Initiation models of r–R pairing via TAR and cTAR. TAR–cTAR duplex formation is initiated either from the bottom of stem-loops that are destabilized by NC (zipper pathway) or from apical loops that are accessible (kissing pathway). The initial interactions limited to a few base pairs require the action of the NC to expand and form the stable TAR–cTAR duplex. For a color version of this figure, see www.iste.co.uk/fosse/structures.zip*

Based on the studies mentioned above and performed with isolated TAR and cTAR stem-loops, two models of r–R pairing initiation have been proposed (see Figure 2.9). In both models, NC is absolutely required to convert the initiation complexes into the TAR–cTAR duplex. In one model, initiation occurs at the bottom of the stems (zipper pathway), whereas in the other model, it occurs at the apical loops (kissing pathway). One of the two initiation pathways is not favored when r–R pairing is performed in vitro with the whole ssDNA and the 3' end of the gRNA; a third pathway is thought to involve the poly(A) and cpoly(A) stem-loops (Chen et al. 2016b). There are no studies that have been able to determine how ex vivo/in vivo r–R pairing is initiated.

2.4. RNA structures promoting genetic recombination

Retroviral recombination is implicated in retrovirus escape from the host immune system and the development of multiple resistance to antiretroviral agents (Simon-Loriere et al. 2011). There are between 5 and 14 recombination events per genome during an HIV-1 replicative cycle (Cromer et al. 2016). These events occur when two gRNA molecules that are genetically different but similar have been packaged into the same viral particle as a heterodimer. Retroviral recombination depends primarily on the internal strand transfer that occurs during reverse transcription; more precisely, the synthesis of the negative polarity DNA strand (see Figure 2.2, step 3). A recombination event is the result of an internal strand transfer event.

2.4.1. *Internal strand transfer*

The molecular mechanisms governing internal strand transfer are similar to those involved in the first strand transfer. Specifically, the RNA template, corresponding to one strand of the heterodimer, must be degraded by the RNase H activity of RT after being copied by this enzyme in order to allow the release of the newly synthesized DNA and its pairing with the acceptor RNA corresponding to the other strand of the heterodimer (see Figure 2.10). Five to fourteen strand transfer events can occur along the gRNA during an HIV-1 replication cycle, because the two strands of gRNA constituting the heterodimer are similar and therefore contain long identical regions where base-pairing of the newly synthesized DNA with the acceptor RNA is possible. Internal strand transfer has been studied mainly in vitro, as it is difficult to characterize ex vivo/in vivo molecular mechanisms (Basu et al. 2008). Simplified experimental systems using RT, two in vitro synthesized RNAs and a DNA primer have been used to reproduce in vitro internal strand transfer (Derebail et al. 2003). Under these experimental conditions, NC strongly increases internal strand transfer in regions where the RNA is highly structured. In contrast, the stimulatory effect of NC on internal strand transfer is low in weakly structured RNA regions.

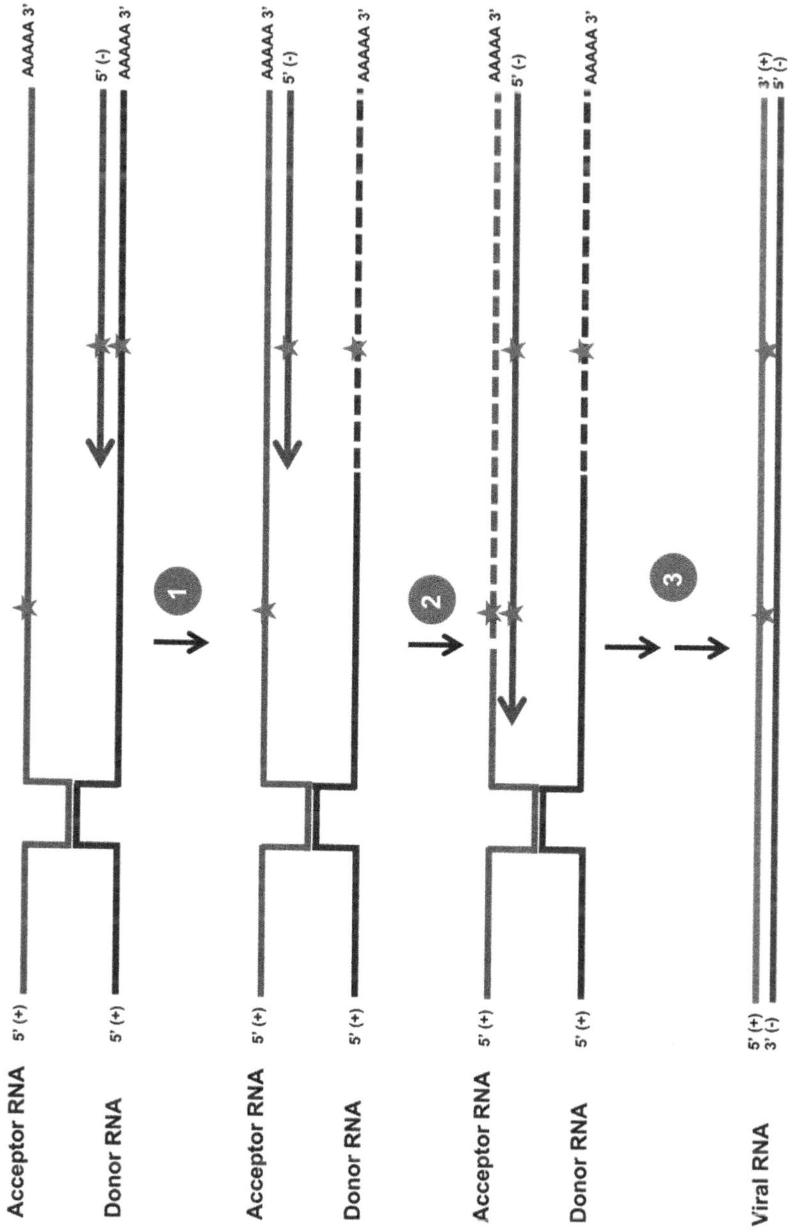

Figure 2.10. *Internal strand transfer. For a color version of this figure, see www.iste.co.uk/fosse/structures.zip*

COMMENTARY ON FIGURE 2.10. – *Example of the acquisition of resistance to two antiretrovirals by internal strand transfer. The gRNA dimer consisting of two RNA strands of positive polarity is shown as two lines colored purple (acceptor RNA) and blue (donor RNA). The region that associates the two subunits of the dimer is shown by the two juxtaposed lines. The red star on the donor RNA corresponds to a mutation conferring resistance to an antiretroviral targeting RT, and the one on the acceptor RNA corresponds to a mutation conferring resistance to an antiretroviral targeting the viral protease. The DNA strand of negative polarity being synthesized is represented by the brown arrow. The synthesis of this strand starts on the donor RNA strand and thus contains the copy of the mutation conferring resistance to an antiretroviral targeting RT. (1) After degradation by the RNase H activity of the RT of the part of the RNA that has just been copied (dashed line), internal strand transfer occurs by pairing of the newly synthesized DNA with the complementary part of the acceptor RNA. (2) DNA strand synthesis continues, and the strand contains a copy of the mutation conferring resistance to a protease-targeting antiretroviral. (3) Subsequent steps in reverse transcription lead to double-stranded viral DNA containing both mutations.*

2.4.2. RNA structures triggering internal strand transfer

Several in vitro studies have shown that the structure of the acceptor RNA plays an important role in internal strand transfer (Negroni and Buc 2000; Moumen et al. 2003; Hanson et al. 2005). The transfer reaction depends generally on a stem-loop present in the acceptor RNA (Moumen et al. 2003; Galetto et al. 2004). Depending on the acceptor RNA, the crossover point, that is, the site where the RT starts to copy the acceptor RNA, is located in or near the stem-loop (Roda et al. 2002; Balakrishnan et al. 2003; Moumen et al. 2003). Depending on the template RNA, the NC changes (Negroni and Buc 2000; Hanson et al. 2005) or does not change (Roda et al. 2002; Balakrishnan et al. 2003) the crossover point. Internal strand transfer is also facilitated by RNA structures resulting from interactions between the donor and acceptor strands. These structures bringing the two strands together can be two stem-loops that interact via their apical loop or tetrads of guanines (Balakrishnan et al. 2003; Shen et al. 2011).

The role of acceptor RNA structure in internal strand transfer is incorporated into the acceptor invasion model (Negroni and Buc 2000) that is consistent with the results of many studies (see references cited by Basu et al. (2008)). In this model (see Figure 2.11), degradation of a portion of the donor RNA by the RNase H activity of RT creates an invasion site where a single-stranded DNA region can interact with the acceptor RNA. This invasion step corresponds to the first base-pairing interaction between the newly synthesized DNA and the acceptor RNA that occurs upstream of the crossover point. This initial interaction expands by displacing the donor RNA fragments, which had remained paired to the DNA. The template exchange occurs during DNA polymerization when the 3' end of DNA pairs with the acceptor RNA. Thanks to its nucleic acid chaperone activity, NC facilitates the above-mentioned molecular mechanisms and thus stimulates internal strand transfer.

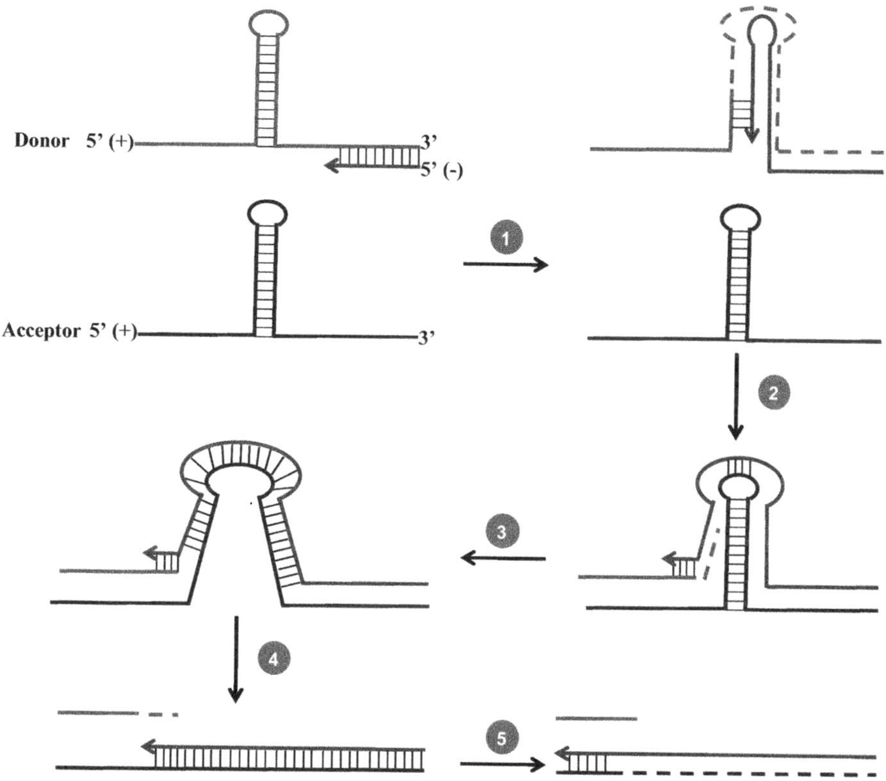

Figure 2.11. *Acceptor invasion model. For a color version of this figure, see www.istc.co.uk/fosse/structures.zip*

COMMENTARY ON FIGURE 2.11. – *Internal strand transfer is triggered by the stem-loop present in the acceptor RNA. The acceptor and donor RNA strands are in blue and purple, respectively. Dashed lines indicate the regions degraded by RT RNase H activity. The negative polarity DNA being synthesized is represented by the brown arrow. Step 1, the RT copies the donor strand and degrades the copied part through its RNase H activity. Step 2, in this example, the acceptor invasion site corresponds to the apical loop, the nucleotides of which are accessible and can pair with the complementary part of the newly synthesized DNA that is released from the donor strand by the RNase H activity of the RT and the NC. Step 3, the initial DNA–acceptor RNA interaction expands and is facilitated by the NC while the RT continues to copy the donor RNA. Step 4, the 3' end of the DNA pairs with the complementary portion of the acceptor strand. Step 5, DNA synthesis continues on the acceptor strand, the copied portion of which is in turn degraded by the RNase H activity of the RT.*

3

RNA Structures Regulating the Expression of the Retroviral Genome

3.1. Regulatory RNA structures of proviral DNA transcription

3.1.1. *The transcriptional activator Tat*

Among the different retroviral species (see Table 1.1), those belonging to the genus lentivirus are characterized by, among other things, their ability to produce the regulatory protein transactivator of transcription (Tat), which is a transcriptional activator. The activation of the transcription of proviral DNA by Tat is, depending on the lentivirus, independent of or dependent on an RNA structure called the transactivator response element (TAR). The lentiviruses bovine immunodeficiency virus (BIV), equine infectious anemia virus (EIAV), Jembrana disease virus (JDV), simian immunodeficiency virus (SIV), HIV-1 and HIV-2 belong to the group where transcriptional activation is dependent on Tat binding to TAR (Ott et al. 2011).

Despite their small size (75–101 amino acids), Tat proteins, interacting with TAR structures, have two essential domains that are conserved among different lentiviruses (see Figure 3.1). The transactivation domain is cysteine-rich and has a highly conserved motif called the core. This motif plays a crucial role in the transactivating function of Tat proteins, as it is directly involved in the interaction with the positive elongation factor P-TEFb. The RNA-binding domain contains an arginine-rich basic motif, which confers its function. In addition, this motif plays an important role in Tat's stability and nuclear localization. Tat proteins are inherently unstructured and structure themselves by interacting with their ligands (Ott et al. 2011; Schulze-Gahmen and Hurley 2018).

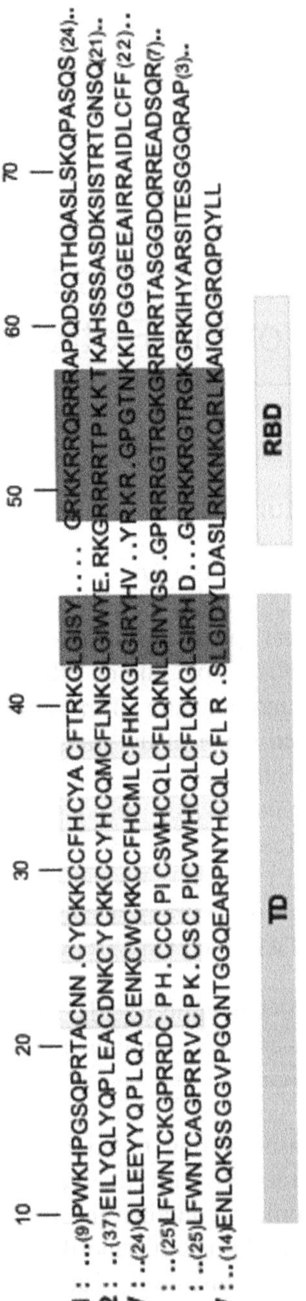

Figure 3.1. *Alignment of six Tat protein sequences. For a color version of this figure, see www.iste.co.uk/fosse/structures.zip*

COMMENTARY ON FIGURE 3.1.– *The numbers in parentheses indicate the numbers of amino acids that are present in the N- and C-terminal ends but not represented in this alignment. Above the HIV-1 Tat sequence are the amino acid positions. Conserved cysteines and histidines are highlighted in yellow. Conserved core amino acids are highlighted in purple. The arginine-rich motif is highlighted in blue. TD indicates the transactivation domain. RBD indicates the RNA binding domain.*

In lentiviruses whose activation of transcription depends on the Tat–TAR interaction, the molecular mechanisms governing the latter are similar but not identical. In all of them, the TAR structure is formed by a part of the R sequence located at the 5' end of the genome, but it can be formed by one or two stem-loops.

3.1.2. *TAR structures with one stem-loop*

BIV, EIAV, JDV and HIV-1 lentiviruses possess a single stem-loop TAR structure (Feng and Holland 1988; Carpenter et al. 1993; Chen et al. 1999; Anand et al. 2008). With the exception of EIAV, a bulge composed of one to three nucleotides plays an essential role in Tat protein binding to TAR RNA. The TAR stem-loop is relatively long in HIV-1 (between 53 and 55 nucleotides depending on the different isolates), whereas it is rather short (between 25 and 28 nucleotides) in BIV, EIAV and JDV lentiviruses (see Figure 3.2). The apical loop and the bulge of the HIV-1 TAR stem-loop are essential for the activation of transcription. In contrast, the lower part of the stem-loop is not required.

3.1.2.1. *TAR–Tat interaction in HIV-1*

The activation of proviral DNA transcription by the Tat–TAR interaction has been most extensively studied in HIV-1. In the absence of this interaction, HIV-1 genome expression can only be very weak and produces mainly short RNAs resulting from transcripts that are stopped at the elongation step (Ott et al. 2011). A positive feedback loop regulates Tat protein production. Specifically, the initial synthesis of Tat resulting from very low-level transcription in the absence of this protein will allow the protein to activate transcription of the viral genome, and thus iteratively increase Tat production and viral particle production (Ott et al. 2011).

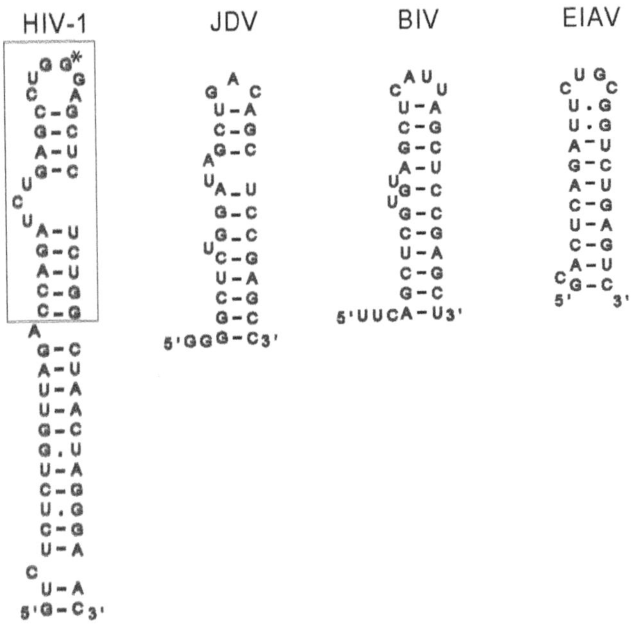

Figure 3.2. *TAR structures with one stem-loop*

COMMENTARY ON FIGURE 3.2. – *Each TAR stem-loop is formed by the 5′ end of the R sequence of the viral genome. The HIV-1 TAR stem-loop shown corresponds to that of the NL4-3 isolate with a three-nucleotide bulge; in some isolates, the bulge is two nucleotides. Although the HIV-1 TAR stem-loop is much longer than other stem-loops, it is only the top half (boxed portion) that is involved in transcription activation. The asterisk denotes the guanine that interacts with the Tat protein loop in the Tat–SEC–TAR complex.*

In the absence of Tat, and in the majority of cases, the transcription of proviral DNA is not continued beyond the initiation step (see Figure 3.3). The elongation step of transcription depends on Tat's recognition of the bulge formed by the TAR stem-loop. The arginine-rich motif of Tat interacts with the nucleotides constituting the bulge, and also those nearby that form the major groove of the RNA double helix (Schulze-Gahmen and Hurley 2018; Chavali et al. 2019). The affinity of Tat for the TAR stem-loop is low and the Tat–TAR interaction is not sufficient to activate proviral DNA transcription, which is a complex process. Indeed, Tat must recruit the super elongation complex (SEC) consisting of the transcription factors ENL/AF9 and ELL2, AFF4 protein, cyclin T1 and CDK9 kinase (Schulze-Gahmen et al. 2016). Cyclin T1 and Tat each interact with AFF4 (see Figure 3.3). The transactivation domain of Tat (see Figure 3.1) directly interacts with cyclin T1. The positive elongation factor P-TEFb resulting from the cyclin T1–CDK9 association inherently lacks the ability to bind to TAR (Romani et al. 2010). In contrast, the affinity of the Tat–P-TEFb complex for TAR is higher than that of Tat alone because cyclin T1 interacts with the apical loop (see Figure 3.3). In addition, the Tat–P-TEFb complex has its affinity increased for the TAR stem-loop when associated with AFF4 (Schulze-Gahmen et al. 2016). In the Tat–SEC–TAR complex, the Tat loop, which is composed of amino acids N24–Y26 (see Figure 3.1) and results from zinc coordination by cysteines, directly interacts via hydrogen bonds with a guanine present in the apical loop of the TAR RNA (see Figure 3.2) (Schulze-Gahmen and Hurley 2018). Formation of the Tat–SEC–TAR complex stimulates the elongation step of transcription by allowing CDK9 to phosphorylate the C-terminus of RNA polymerase II and the negative elongation factors NELF and DSIF.

3.1.2.2. *TAR–Tat interaction in SIVcpz, BIV, JDV and EIAV lentiviruses*

The activation of transcription by TAR–Tat interaction in SIVcpz (SIV infecting chimpanzees) is similar to that observed in HIV-1 (D'Orso and Frankel 2009). In contrast, in bovine lentiviruses (BIV and JDV), the activation of transcription is different because P-TEFb has the primary role of phosphorylating the c-terminus of RNA polymerase II. More specifically, cyclin T1 does not interact with the TAR stem-loop, and Tat has a high affinity for it that is independent of its interaction with cyclin T1 (D'Orso and Frankel 2009). Tat does not interact with the apical loop of TAR but only with the major RNA groove located near the central bulge, corresponding to UGU in BIV and UA in JDV (see Figure 3.2).

In EIAV, the activation of transcription also requires the formation of the P-TEFb–Tat complex, which depends on the insertion of the C-terminal leucine of Tat (see Figure 3.1) into the hydrophobic pocket formed by two helices of cyclin T1 (Anand et al. 2008). The arginine-rich motif of Tat and adjacent amino acids interact with the apical loop and the upper part of the

TAR stem-loop. Cyclin T1 increases the affinity of the P-TEFb–Tat complex for TAR because it interacts with part of the apical loop (Anand et al. 2008).

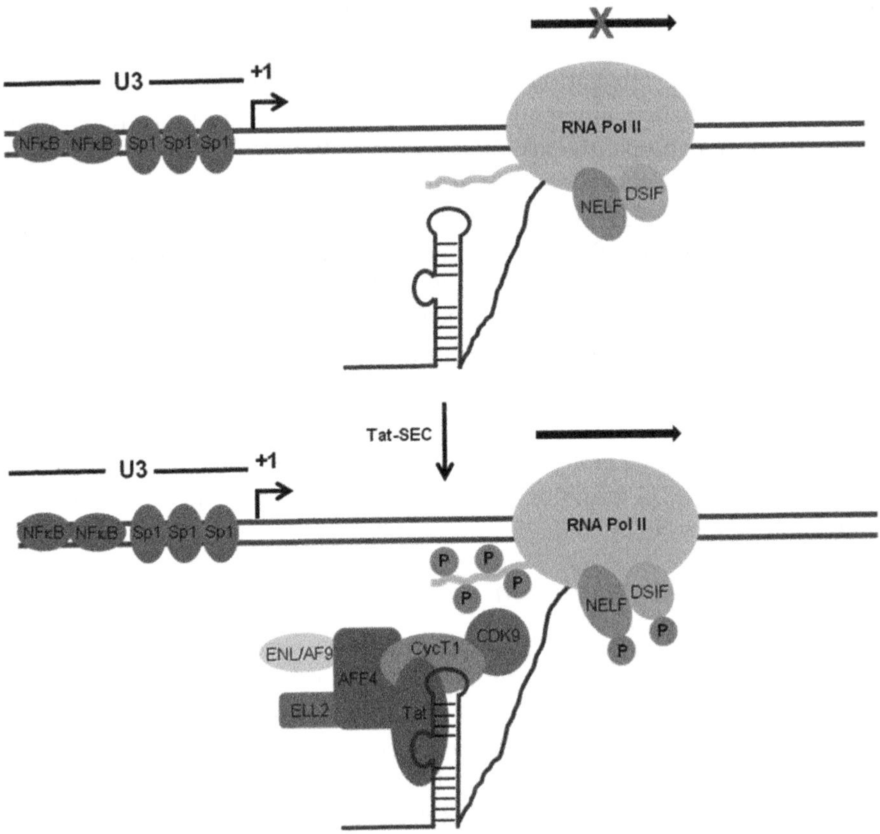

Figure 3.3. *Activation of transcription by the HIV-1 Tat protein. For a color version of this figure, see www.iste.co.uk/fosse/structures.zip*

COMMENT ON FIGURE 3.3.– *In the absence of Tat, cellular transcription factors (e.g. NF-κB and Sp1) allow the initiation of transcription by RNA polymerase II by binding to the U3 portion of the 5' LTR. This initiation step, which is not followed by an elongation step (shown by an arrow with a red cross), generates short transcripts (50–60 nucleotides) that can form the upper part of the TAR stem-loop. Elongation by the polymerase is inhibited by the protein factors NELF and DSIF. The elongation step requires that Tat recruits the super elongation complex (SEC) which consists of the proteins cyclin T1 (CycT1), CDK9, AFF4, ENL/AF9 and ELL2. Tat and CycT1 are the two proteins in the Tat–SEC complex that bind it to TAR RNA. Tat interacts with the bulge and part of the apical loop, whereas CycT1 interacts with the apical loop. The Tat–SEC–TAR interaction leads CDK9 to phosphorylate NELF and DSIF and thus counteract their inhibitory effect on elongation. CDK9 hyperphosphorylation of the C-terminus of RNA polymerase II leads the latter to perform elongation.*

3.1.3. *TAR structures with two stem-loops*

SIVmac (SIV infecting macaques) and HIV-2 are representative of lentiviruses with a two stem-loop TAR structure (D'Orso and Frankel 2009) (see Figure 3.4). Each stem-loop has a central two-nucleotide bulge (UU and UA). The stem-loop with the UU bulge appears to play a more important role than the other stem-loop in activating transcription (Fenrick et al. 1989; Rhim and Rice 1994). Transcription activation also depends on the formation of the P-TEFb–Tat complex (D'Orso and Frankel 2009).

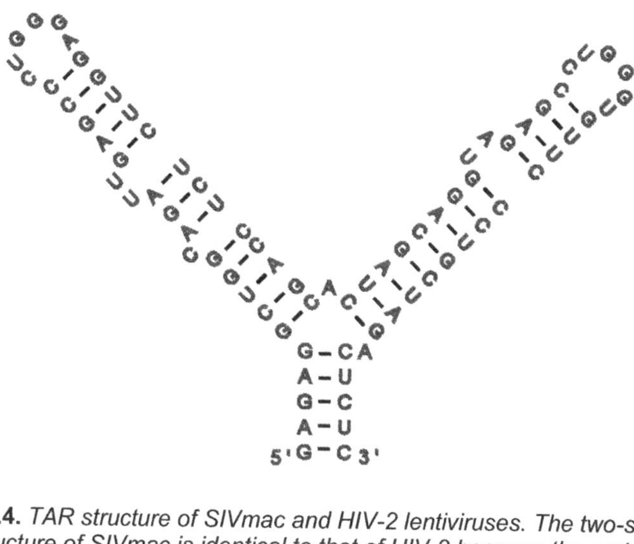

Figure 3.4. *TAR structure of SIVmac and HIV-2 lentiviruses. The two-stem-loop TAR structure of SIVmac is identical to that of HIV-2 because the part of the R sequence that adopts this structure is the same in these two lentiviruses*

Each of the two stem-loops of the TAR structure binds a P-TEFb–Tat complex via an interaction involving Tat and the central bulge. Unlike HIV-1, extensive structural studies of P-TEFb–Tat–TAR complexes have not been performed with Tat proteins and TAR RNAs from SIVmac and HIV-2 lentiviruses.

3.2. RNA structures regulating genomic RNA maturation

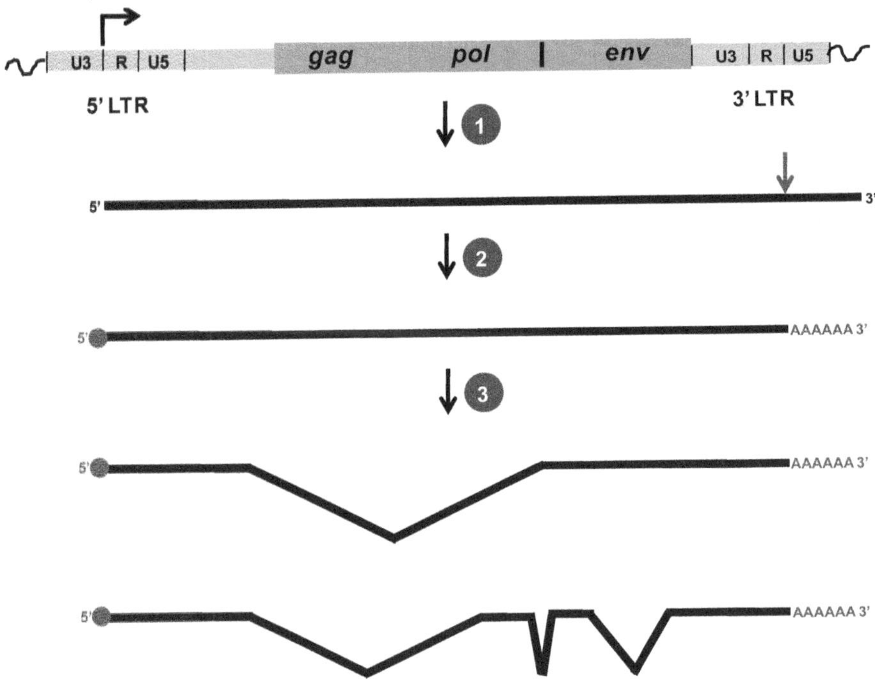

Figure 3.5. *Maturation of retroviral RNA. For a color version of this figure, see www.iste.co.uk/fosse/structures.zip*

COMMENTARY ON FIGURE 3.5.– *1) Proviral DNA is transcribed by RNA polymerase II. 2) Primary structure of the pre-messenger RNA resulting from transcription and initial post-transcriptional modifications: addition of a 5′ cap, cleavage of the 3′ end and synthesis of a poly(A) tail. 3) Splicing: a single event producing, for example, the env mRNA in ALV alpharetrovirus or three events producing, for example, the nef mRNA in HIV-1 lentivirus. The start of transcription is indicated by a horizontal black arrow. The orange arrow indicates the 3′ cleavage. The orange disk represents the cap.*

Retroviral RNAs, like all cellular pre-messenger RNAs synthesized by RNA polymerase II, are the products of rapid maturation leading to the addition of a cap at their 5' end and cleavage of their 3' end followed by the addition of a poly(A) tail consisting of approximately 200 adenosine monophosphate residues (see Figure 3.5). Maturation continues with one or two splicing events in simple retroviruses (e.g. ALV and RSV alpharetroviruses) and multiple splicing events in complex retroviruses (e.g. HIV-1 and HIV-2 lentiviruses). A characteristic of retroviruses is that the majority of capped and polyadenylated RNA molecules are not spliced (McNally 2008). In effect, the unspliced RNA serves as single-stranded genomic RNA which is converted to double-stranded DNA by RT or as mRNA for the synthesis of Gag and Gag–Pol polypeptide precursors. Retroviruses have therefore evolved strategies to control the proportions of spliced and unspliced mRNAs. Cis-regulatory elements are involved in this control (Cochrane et al. 2006). The cis-regulatory properties of some are related to the RNA structures they form. These structures have only been well-characterized in RSV and HIV-1.

3.2.1. *The negative regulator of splicing of RSV*

The RSV belonging to the genus alpharetrovirus was one of the first retroviruses to be studied because, in addition to the *gag*, *pol* and *env* genes, it contains the oncogene sarc (*src*) which codes for a tyrosine kinase involved in cell proliferation. In addition, this virus has served as a relatively simple model for studying alternative splicing in retroviruses. Indeed, the unspliced mRNA representing approximately 75% of RSV mRNAs (Hudson et al. 2016) contains a single splice donor site and two acceptor sites to generate *env* and *src* mRNAs (see Figure 3.6).

The low proportion of spliced RNAs (approximately 25%) is primarily due to three negative splicing signals (Cochrane et al. 2006). The first results from the *env* mRNA branch point, which only partially matches the canonical sequence required for efficient splicing (Katz and Skalka 1990). The second is due to degeneration of the polypyrimidine motif constituting part of the acceptor site of the *src* mRNA (Zhang and Stoltzfus 1995). The third corresponds to the negative regulator of splicing (NRS) that reduces the production of *env* and *src* mRNAs (Arrigo and Beemon 1988; Cochrane et al. 2006). This cis-regulatory element also has a very long-range stimulatory effect on polyadenylation of unspliced mRNA (Hudson et al. 2016). The NRS consists of approximately 230 nucleotides and is located in the *gag* gene (see Figure 3.6).

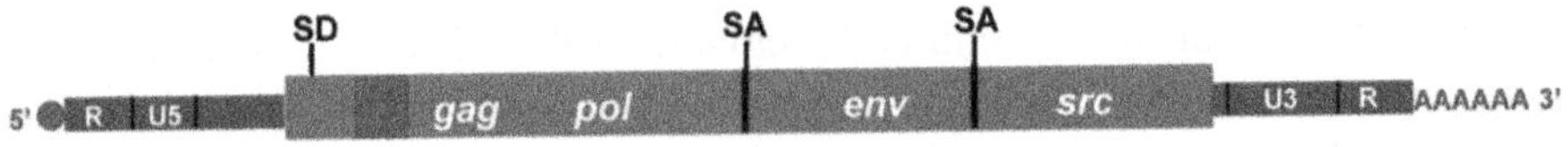

Figure 3.6. *RSV unspliced mRNA. The RSV unspliced mRNA contains one splice donor site (SD site) and two acceptor sites (SA sites) that generate the env and src mRNAs. The NRS sequence, which is colored green, is more than 4,000 nucleotides from the first SA. The orange disk represents the 5' cap. Regions and genes are not shown to scale. For a color version of this figure, see www.iste.co.uk/fosse/structures.zip*

The secondary structure of NRS was determined using chemical and enzymatic probes (Bar et al. 2011). The 5' and 3' ends of the NRS interact to form a cruciform structure that is conserved among different alpharetroviruses (see Figure 3.7).

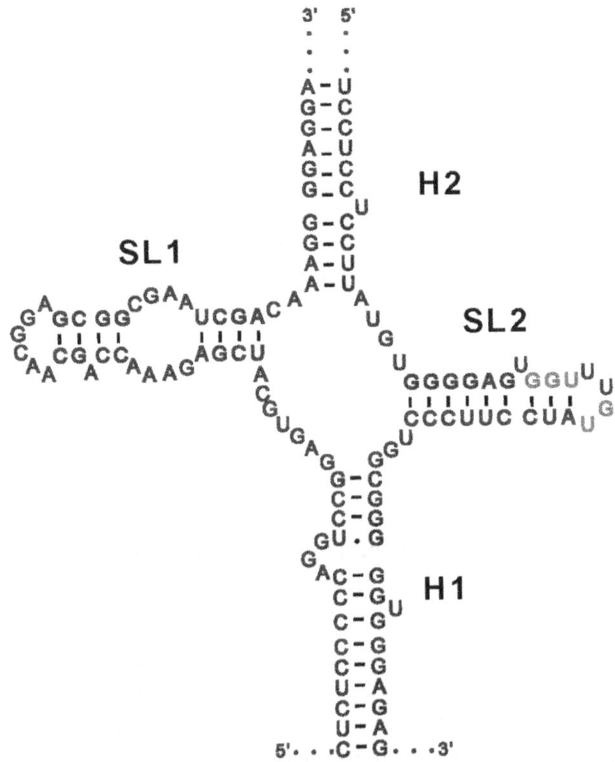

Figure 3.7. *Cruciform structure of the NRS. For a color version of this figure, see www.iste.co.uk/fosse/structures.zip*

COMMENTARY ON FIGURE 3.7.– *The 5' and 3' ends of NRS interact to form a cruciform structure that is responsible for the splicing inhibitory activity of NRS. Helices 1 and 2 are called H1 and H2, respectively. Stem-loops 1 and 2 are called SL1 and SL2, respectively. The apical part of SL2 partially corresponds to the splice donor site consensus sequence (AGGURAGU, R being a purine). The orange-colored nucleotides are complementary to the U1 small RNA region that matches the splice donor consensus sequence.*

The cruciform structure is responsible for the splicing inhibitory effect of NRS. The hnRNP H and SR proteins (ASF/SF2, 9G8 and SC35), which are splicing-regulating factors, bind to the cruciform structure. The splicing inhibitory activity of NRS has not been shown to be dependent on the binding of hnRNP H proteins. Stem-loop 1 is the major binding site for the SR proteins ASF/SF2 and 9G8 (McNally and McNally 1996; Bar et al. 2011). The H2 helix also binds the SR 9G8 protein, whereas the H1 helix and stem-loop 2 are not binding sites for this protein. The activity of NRS is related, at least in part, to its interaction with the SR 9G8 protein. The results obtained with SR proteins led to the hypothesis that SR proteins recruit the nuclear ribonucleoprotein U1, which recognizes a pseudo-splice donor site in NRS. Indeed, the apical part of the stem-loop 2 has a pseudo-donor site that enables the binding of the small nuclear RNA U1 that participates in splicing (see Figure 3.7). The conformation of stem-loop 2 must not be altered in order to bind the U1 RNA (Cabello-Villegas et al. 2004). This binding leads to the formation of a non-functional spliceosome, in which the pseudo-donor site is coupled with one of the 3' acceptor sites. This coupling prevents the 3' acceptor site from interacting with the authentic 5' donor site. It has also been hypothesized that the SR proteins are required for the coupling of the pseudo-donor site to one of the 3' acceptor sites. Furthermore, by binding to the NRS, SR proteins allow the NRS to stimulate 3' polyadenylation of the newly synthesized RNA (Hudson et al. 2016).

3.2.2. *Structural diversity and alternative splicing in HIV-1*

HIV-1 pre-messenger RNA (~9,200 nucleotides) undergoes extensive alternative splicing that produces more than 50 mRNA isoforms, allowing for optimized expression of viral proteins required for viral replication (Sertznig et al. 2018). This is due to the presence of at least four donor sites and eight acceptor sites in all HIV-1 isolates. The many possibilities of alternative splicing were illustrated by the identification of 109 spliced RNAs when analyzing a clinical HIV-1 isolate with massive parallel sequencing (Ocwieja et al. 2012). In this particular case, the pre-messenger RNA contains 7 donor sites and 14 acceptor sites. Alternative splicing results from various complex mechanisms. One of these mechanisms relies on the strength of a donor or acceptor site, which depends on the degree of similarity to the consensus sequence (Sertznig et al. 2018). The greater the difference between the site sequence and the consensus sequence, the lower its strength, as well as the amount of mRNA resulting from splicing at that site. Another mechanism involves cis-regulatory elements that, by binding proteins, activate or inhibit splicing at a given site. Splicing at a site may depend on a combination of these two mechanisms.

A third recently discovered mechanism relies on the regulation of alternative splicing by the structural diversity of HIV-1 pre-messenger RNA. Indeed, pre-messenger RNA does not adopt a single conformation but several alternative conformations ex vivo (Tomezsko et al. 2020). Pre-messenger RNA molecules are therefore present in different conformations in the nucleus of the infected cell. Depending on the conformation adopted by a pre-messenger RNA molecule, splicing at a given site can be repressed or activated. The structural diversity of HIV-1 pre-messenger RNA partly explains its alternative splicing and the persistence of unspliced pre-messenger RNA, which acts both as genomic RNA and as mRNA for the *gag* and *pol* genes. The role of RNA conformation was best illustrated by studying the A3 acceptor site, which is required for the production of Tat mRNAs. The regulation of splicing by the RNA structure at the A3 site can reduce the amount of Tat mRNAs by a factor of 100 (Tomezsko et al. 2020). Specifically, the region of the pre-messenger RNA containing the A3 site forms two distinct and major secondary structures (see Figure 3.8). In one (structure 1), present in 63–67% of pre-messenger RNA molecules, the pyrimidine-rich sequence of the A3 site is fully accessible and can therefore bind the U2AF protein that plays a crucial role in 3' splicing. In contrast, in the other secondary structure (structure 2), present in 33–37% of pre-messenger RNA molecules, this sequence is inaccessible to U2AF because it is largely engaged in intramolecular pairing.

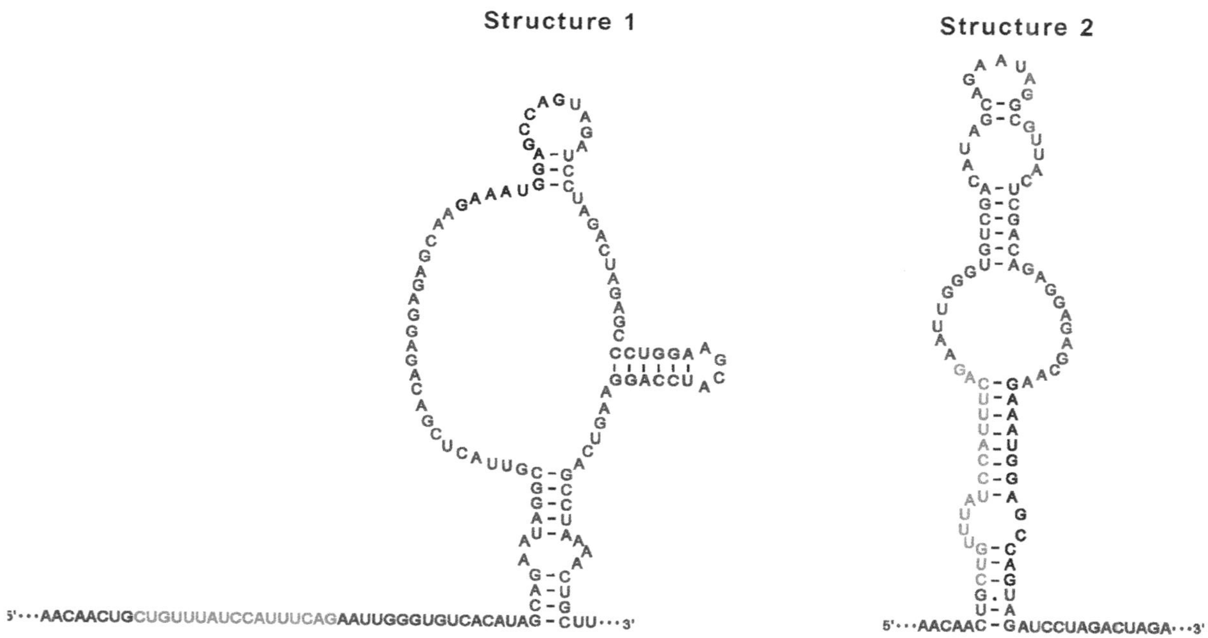

Figure 3.8. *Secondary structures of the A3 site-containing region. The HIV-1 pre-messenger RNA region containing the A3 site can adopt two different conformations. The A3 site (shown in orange) containing the pyrimidine-rich sequence is accessible in structure 1 but is not accessible in structure 2 because it is almost completely paired with another sequence (black nucleotides). For a color version of this figure, see www.iste.co.uk/fosse/structures.zip*

3.3. RNA structures regulating the export of retroviral RNAs

In eukaryotes, almost all cellular pre-messenger RNAs undergo splicing to remove introns, and only spliced mRNAs can reach the cytoplasm for translation. Nuclear retention of unspliced or partially spliced transcripts prevents the synthesis of non-functional or cell-deleterious proteins. In contrast, the majority of retroviral pre-messenger RNAs must be able to be exported to the cytoplasm because they have two functions:

– mRNA for the synthesis of Gag and Gag–Pol polypeptide precursors;

– genomic RNA encapsidated in the viral particle.

As a result, retroviruses have evolved strategies to circumvent cellular restrictions that prevent RNAs containing one or more introns from being transported from the nucleus to the cytoplasm (Shida 2012). The export mechanisms of unspliced retroviral RNAs can be divided into two groups corresponding to complex and simple retroviruses.

3.3.1. *Export of unspliced RNA in simple retroviruses*

The export mechanisms of unspliced retroviral RNA in simple retroviruses are different from those used by complex retroviruses. Indeed, the transport of unspliced RNA from the nucleus to the cytoplasm does not require a regulatory protein encoded by the retrovirus. Moreover, the NXF1–NXT1 pathway (also called Tap–p15) is involved in the nuclear export of unspliced retroviral RNA. This pathway, requiring the formation of a heterodimer between the NXF1 and NXT1 proteins, is also used for the transport of cellular mRNAs from the nucleus to the cytoplasm (Katahira 2015). NXF1 is a medium-sized cellular protein (619 amino acids) that has several functional domains (see Figure 3.9). One, located in the N-terminal part, corresponding to a nuclear localization signal (NLS), allows NXF1 to be imported into the nucleus. The N-terminal half of NXF1 contains an RNA recognition domain (RRM) and a leucine-rich domain (LRR). The binding of the NXF1–NXT1 heterodimer to RNA involves the RRM, LRR and NTF2L domains (Katahira 2015). The binding of NXF1–NXT1 to most cellular mRNAs is not direct; it requires interaction with the TREX1 protein complex beforehand (Viphakone et al. 2012). The C-terminal half of NXF1, allowing interaction with the nuclear pore complex (NPC), contains the NTF2L and UBA domains. NXF1 associates with NXT1 via the NTF2L domain.

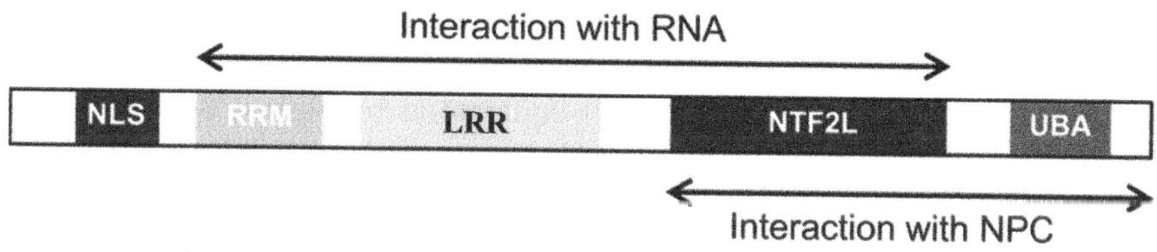

Figure 3.9. *Structural organization of NXF1. NXF1 has five domains. The domains are not shown to scale. For a color version of this figure, see www.iste.co.uk/fosse/structures.zip*

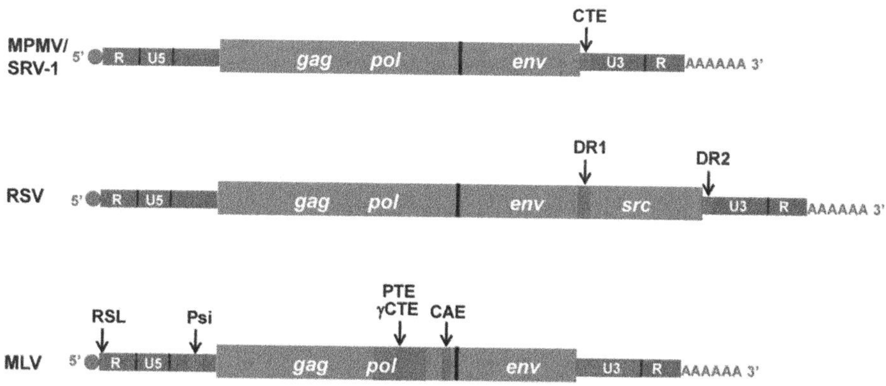

Figure 3.10. *Unspliced mRNAs of three simple retroviruses. For a color version of this figure, see www.iste.co.uk/fosse/structures.zip*

COMMENTARY ON FIGURE 3.10.– *Unspliced mRNAs of three simple retroviruses belonging to three different genera are shown as examples (MPMV and SRV-1 for simple betaretroviruses, RSV for alpharetroviruses, MLV for gammaretroviruses). The cis-regulatory regions of export are in green and indicated by arrows, except for the one that is present in the U3 region of MLV but has not been precisely localized. The γCTE element is included in the PTE region of MLV. Regions and genes are not shown to scale. The cap is indicated by an orange disk.*

In simple retroviruses, in contrast to complex retroviruses, several regions of the unspliced RNA may be involved in its transport (see Figure 3.10). These cis-regulatory regions may consist of single or multiple stem-loops. The protein–RNA interactions involved in the export of unspliced RNAs from simple retroviruses have not yet been characterized or are only partially characterized. Nuclear export of unspliced RNAs from MPMV and SRV-1, two betaretroviruses, is relatively well-understood because it depends on a single cis-regulatory region (CTE) and the single NXF1–NXT1 pathway. Nuclear export is more complex in RSV, an alpharetrovirus, because it depends on two repeat sequences (DR1 and DR2) and, in addition to the NXF1–NXT1 pathway, the Crm1 pathway could be involved via an interaction with the viral polypeptide Gag (this point will be discussed in Chapter 4). The most complex nuclear export is observed in MLV, a gammaretrovirus, as it involves five regions and also the Crm1 pathway in addition to the NXF1–NXT1 pathway (Pessel-Vivares et al. 2015; Mougel et al. 2020).

3.3.1.1. *Export of unspliced MPMV and SRV-1 RNAs*

Nuclear export of unspliced RNAs from MPMV and SRV-1 retroviruses depends on the CTE element consisting of 154 nucleotides (Bray et al. 1994; Zolotukhin et al. 1994). It is likely that the CTE element is of cellular origin and was captured by a common ancestor of MPMV and SRV-1 (Wang et al. 2015). Specifically, intron 10 of the primate NXF1 gene contains a region of approximately 100 nucleotides that has high sequence and secondary-structure homology to the MPMV CTE (Li et al. 2006). This region allows the nuclear export of NXF1 mRNAs containing intron 10. An in vitro study using chemical and enzymatic probes showed that the MPMV CTE region folds into a long stem-loop (Ernst et al. 1997). This stem-loop is characterized by two inner loops that are arranged in mirror symmetry (see Figure 3.11).

The use of site-directed mutagenesis has demonstrated that nuclear export of unspliced RNA is not possible when both inner loops of the CTE of SRV-1 or MPMV are deleted (Pasquinelli et al. 1997; Grüter et al. 1998). These studies also showed that the CTE remains functional when it has only one inner loop. However, the two loops are not equivalent; loop A is more efficient than loop B for nuclear export. Duplication of the inner loop in the CTE likely optimizes the export of unspliced RNA (Wang et al. 2015).

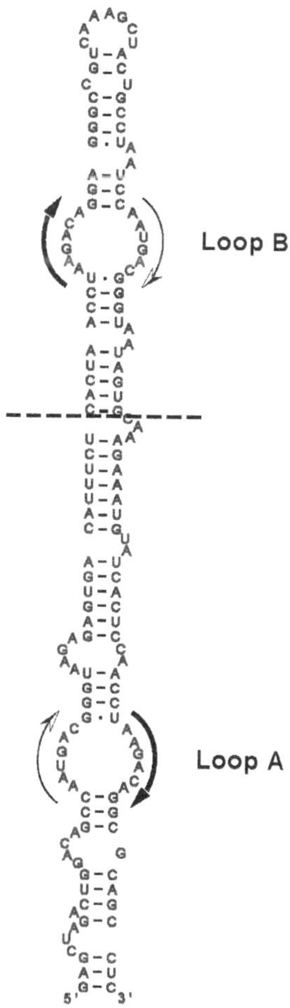

Figure 3.11. *Secondary structure of the CTE RNA of MPMV. The CTE element forms a single stem-loop. Arrows indicate identical sequences. The portion above the dashed line corresponds to the top part of the CTE called hCTE*

NXF1 directly binds to each inner loop of the CTE (Grüter et al. 1998; Kang and Cullen 1999; Teplova et al. 2011). The truncated form of NXF1, restricted to the RRM and LLR domains (see Figure 3.9), can bind to the entire CTE or its upper part called hCTE (see Figure 3.11) containing the inner loop B (Braun et al.1999; Kang and Cullen 1999; Teplova et al. 2011). The three-dimensional structure of the truncated NXF1–hCTE complex can be determined by X-ray crystallography (Teplova et al. 2011). The hCTE RNA adopts an L-shaped conformation when it associates with the truncated NXF1 protein. In this complex, the RRM domain interacts with the ribose-phosphate backbone constituting the inner loop and the double-stranded segments that flank it. The LRR domain interacts with bases and the ribose-phosphate backbone that constitute the single-stranded regions of the inner loop and the AA bulge. The truncated NXF1–hCTE complex only partially accounts for the interactions between the whole CTE and the two NXF1–NXT1 heterodimers. Indeed, the NTF2L domain of NXF1 interacts with RNA when the protein is associated with NXT1 (Katahira 2015). Thus, the RRM, LRR and NTF2L domains of NXF1 are involved in CTE element recognition. To date, there is no three-dimensional structure of the complex formed by the whole CTE and the two NXF1–NXT1 heterodimers.

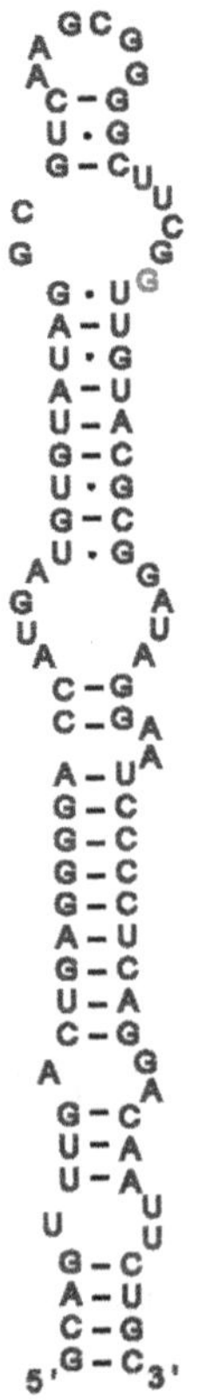

Figure 3.12. *Secondary structure of a portion of the RSV DR2 RNA. A large portion of the RSV DR2 element (Prague C strain) folds into a single stem-loop. The orange-colored guanine plays a crucial role because its replacement by a cytosine strongly inhibits retrovirus replication and nuclear export. For a color version of this figure, see www.iste.co.uk/fosse/structures.zip*

3.3.1.2. *Export of unspliced RNA from RSV*

The export of unspliced RNAs from RSV, an alpharetrovirus, is more complex than that of unspliced RNAs from MPMV. Indeed, RSV unspliced RNAs appear to use two different pathways to be transported from the nucleus to the cytoplasm (Maldonado et al. 2020). Although they are identical, the unspliced RNAs are thought to split into two populations. Each of the unspliced RNAs in one of the populations functions as a genomic RNA by associating with the Gag polypeptide, using the Crm1 pathway to pass into the cytoplasm where it is directed to the plasma membrane and encapsidated into the virus particle (this process is described in Chapter 4). The unspliced RNAs from the other population make use of the NXF1 and Dbp5 (DEAD box RNA helicase) pathway to pass to the cytoplasm, where they serve as mRNA for Gag or Gag–Pol synthesis (LeBlanc et al. 2007). Both RSV DR elements (approximately 135 nucleotides each) can be used in export via the NXF1 pathway, although one is sufficient (Ogert et al. 1996; Simpson et al. 1997). Since the NXF1 protein does not directly bind to DR1 and DR2 (Paca et al. 2000), it may interact with a previously uncharacterized protein that binds to DR elements (LeBlanc et al. 2007).

Although the DR1 and DR2 sequences are functionally interchangeable and show an 82% identity, they do not adopt an identical secondary structure (Paca et al. 2000). There are, however, a few conserved nucleotides in the 3' half of the DR elements of different alpharetrovirus strains that are essential for viral replication and nuclear export (Ogert and Beemon 1998; Yang and Cullen 1999; Paca et al. 2000; LeBlanc et al. 2007). For example, one of these corresponds to a guanine in the long stem-loop formed by a large portion of the DR2 element of the RSV Prague C strain (see Figure 3.12).

3.3.1.3. *Export of unspliced RNA from MLV*

Although MLV is defined as a simple retrovirus, the nuclear export of its unspliced RNAs appears to be the most complex in the retrovirus world. MLV unspliced RNAs use two different pathways for transport from the nucleus to the cytoplasm (Mougel et al. 2020). The unspliced RNAs are divided into two populations according to their function in the cytoplasm. One population corresponds to genomic RNAs that use the Crm1 pathway to pass into the cytoplasm and are then directed to the plasma membrane where they are encapsidated in viral particles (Mougel et al. 2020). As in the case of RSV, the Gag polypeptide could be involved in transport via the Crm1 pathway but this has not been demonstrated. Unspliced RNAs from the other population make use of the NXF1 pathway to be exported to the cytoplasm where they serve as mRNA for Gag or Gag–Pol synthesis (Pessel-Vivares et al. 2014; Sakuma et al. 2014b). The NXF1 protein does not directly bind to the unspliced mRNA. The nuclear export of MLV unspliced mRNAs via the NXF1 pathway involves several proteins of the TREX1 complex, which are the RNA helicase UAP56, THOC5 and THOC7 (Bartels and Luban 2014; Sakuma et al. 2014a). Unlike UAP56 and THOC7, THOC5 is among the proteins strongly associated with unspliced mRNA but has not been shown to directly interact with it. THOC5 may play a role as an adaptor protein by recruiting UAP56 and THOC7 (Sakuma et al. 2014a).

The molecular mechanisms allowing the nuclear export of the unspliced RNA of MLV have not been elucidated because they are complex, and it is difficult to obtain unambiguous results directly linked to an alteration of the nuclear export with site-directed mutagenesis. Indeed, the unspliced RNA, corresponding to mRNA or gRNA, is multi-functional. To circumvent these difficulties, most ex vivo studies have been performed with retroviral vectors which contain only a part of the viral genome, and thus allow for specifically studying certain phases of the replicative cycle of a retrovirus. However, the possibility that an export mechanism characterized with retroviral vectors does not correspond, or only partially corresponds, to that used by the whole viral genome cannot be excluded. Ex vivo studies have identified five cis-regulatory regions of nuclear export of unspliced RNA that are scattered along the viral genome (see Figure 3.10). The cis-regulatory activity present in the U3 region has not been shown to be dependent on one or more RNA structures (Volkova et al. 2014).

The first nuclear export signal was identified using the whole viral genome and retroviral vectors (Trubetskoy et al. 1999). This signal called RSL is located at the 5' end of the unspliced RNA (see Figure 3.10) and its activity is related to its ability to form a stem-loop structure (see Figure 3.13). The proteins that interact with RSL during nuclear export have not been characterized.

Two independent studies using retroviral vectors have shown that the gRNA packaging signal called Psi may also have a role as a cis-regulatory region for nuclear export of unspliced RNA (Basyuk et al. 2005; Smagulova et al. 2005). This region folds in the form of four stem-loops (see Figure 3.13). In the natural context, Psi is probably involved in the nuclear export of unspliced RNA. Indeed, an overexpression of Psi by a retroviral vector in an

MLV-infected cell competitively leads to a strong decrease in the transport of unspliced and whole MLV RNA from the nucleus to the cytoplasm (Smagulova et al. 2005). The proteins involved in this transport have not been identified. The Gag polypeptide that interacts with Psi during gRNA encapsidation (a process described in Chapter 4) could be involved and use the Crm1 pathway but this has not been demonstrated.

A study performed with the whole viral genome and retroviral vectors identified a long region (1,468 nucleotides) in the MLV *pol* gene called PTE (see Figure 3.10) that plays an essential role in the nuclear export of unspliced RNA (Pilkington et al. 2014). The same study determined the secondary structure of the PTE region in vitro using two chemical probes. This folds as seven long stem-loops annotated SL1–SL7 in the 5'–3' direction. SL1 and SL7 were found to play a major role in the activity of the PTE region.

Another study, performed with retroviral vectors, characterized the γCTE element in SL2 (see Figure 3.10), which has similarities to the CTE element of MPMV (Bartels and Luban 2014). Specifically, the AAGACA sequence that is present in the apical loop of γCTE is also present in the NFX1-interacting inner loops of CTE (see Figure 3.13). Although the γCTE element has not been shown to bind the NXF1 protein, the results indicate that it contributes to the nuclear export that depends on this protein. The main function of γCTE is to enable polysome formation during translation of the *gag* and *gag–pol* genes (Bartels and Luban 2014).

A study conducted with retroviral vectors (Sakuma et al. 2014b) identified another nuclear export cis-regulatory element that is also in the *pol* gene, but downstream of the PTE region (see Figure 3.10). This element called CAE can form three short stem-loops (see Figure 3.13). It facilitates the nuclear export of unspliced RNA via the NFX1 pathway. A direct interaction between the NXF1 protein and CAE has not been shown. The CAE element contains the sequence GGAAAGGAC which is highly conserved among gammaretroviruses. This sequence plays an essential role in the CAE activity, but is not sufficient to allow the transport of unspliced RNA from the nucleus to the cytoplasm.

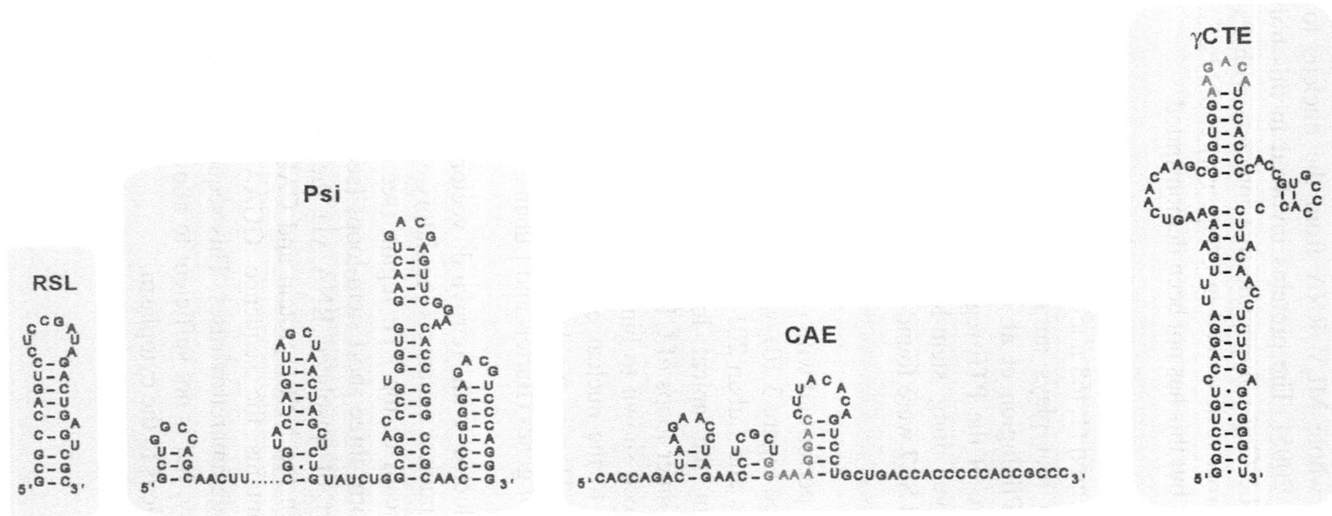

Figure 3.13. *Secondary structures of nuclear export signals in MLV. The sequence in orange in the CAE element is highly conserved among gammaretroviruses. The orange sequence in the γCTE element is also present in the two inner loops of the MPMV CTE. For a color version of this figure, see www.iste.co.uk/fosse/structures.zip.*

3.3.2. *Unspliced RNA export in complex retroviruses*

The mechanisms of unspliced retroviral RNA export are similar in complex retroviruses belonging to the genera betaretrovirus (JSRV, MMTV, etc.), deltaretrovirus (HTLV-1, HTLV-2, etc.) and lentivirus (HIV-1, HIV-2, SIV, etc.). The export of unspliced RNAs, using the Crm1 pathway, depends on a specific interaction between a regulatory protein encoded by the retrovirus and an RNA structure formed by a region of the unspliced retroviral RNA. This cis-regulatory region is located in the 3' part of the genome, and its length and location vary according to the retroviral species (see Figure 3.14). In all complex retroviruses, it adopts a conformation consisting of several stem-loops. The protein–RNA interactions involved in the export of unspliced RNAs from complex betaretroviruses have not been characterized, and those occurring in deltaretroviruses are only partially elucidated. It is in HIV-1 that the export mechanism of unspliced RNA was discovered (Malim et al. 1989) and has been most extensively studied (Kuzembayeva et al. 2014).

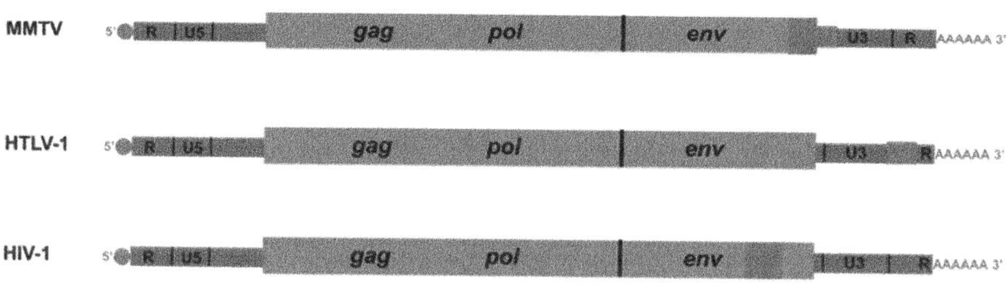

Figure 3.14. *Unspliced mRNAs of three complex retroviruses. For a color version of this figure, see www.iste.co.uk/fosse/structures.zip*

COMMENTARY ON FIGURE 3.14.– *Unspliced mRNAs from three complex retroviruses belonging to three different genera are shown as examples. The cis-regulatory export regions are shown in green and are called RmRE, RxRE and RRE in MMTV, HTLV-1 and HIV-1, respectively. RmRE (490 nucleotides) includes the 3' end of the env gene and the 5' end of the U3 region. RxRE (255 nucleotides) includes the 3' end of the U3 region and a large part of the R sequence. RRE (351 nucleotides) is located in the env gene. Regions and genes are not shown to scale. The cap is represented by an orange disk.*

3.3.2.1. *Export of unspliced HTLV-1 RNA*

In deltaretroviruses (HTLV-1, HTLV-2, etc.), it is the Rex protein that plays an essential role in the export of unspliced or partially spliced viral RNAs (Nakano and Watanabe 2016). Like the Rev protein of lentiviruses, Rex is a regulatory protein that is expressed during the early phase of viral replication. It is translated from doubly spliced viral mRNAs that, like those encoding the Tax protein (equivalent to the HIV-1 Tat protein), are the first to be transported to the cytoplasm after transcription of proviral DNA. These spliced viral mRNAs are transported from the nucleus to the cytoplasm by the same pathway used by cellular mRNAs.

Figure 3.15. *Structural organization of Rex. The N-terminus contains the arginine-rich domain that includes the NLS. The bipartite oligomerization domain and the nuclear export signal (NES) are in the central portion. The C-terminus contains the domain that stabilizes the protein. The domains are not shown to scale. For a color version of this figure, see www.iste.co.uk/fosse/structures.zip*

Rex is a small protein (189 amino acids in HTLV-1) that has several functional domains (see Figure 3.15). One, located in the N-terminal part, allows Rex to be imported into the nucleus and to bind to RNA because it contains an NLS and is rich in arginines (D'Agostino et al. 2019). The central part of Rex contains a bipartite oligomerization domain and an NES. Through NLS and NES signals, Rex can shuttle between the cytoplasm and nucleus (Nakano and Watanabe 2016). The C-terminus corresponds to a domain that stabilizes the protein by increasing its half-life (Xie et al. 2009).

The export of unspliced and partially spliced viral RNAs requires Rex binding to the RxRE region (255 nucleotides) (Nakano and Watanabe 2016). Note that, unlike the HIV-1 cis-regulatory RRE sequence, RxRE is present in all HTLV-1 RNAs because it is located downstream of splice acceptor sites and in the 3' end of the genome (see Figure 3.14). RNAs whose export depends on their interaction with Rex have a 3' CRS element that is responsible for their retention in the nucleus and degradation in the absence of Rex (Cavallari et al. 2016). By binding to RxRE, Rex abolishes the negative effect of CRS by enabling transport from the nucleus to the cytoplasm via the Crm1 pathway. The RxRE RNA forms a secondary structure with four stem-loops (see Figure 3.16) (Toyoshima et al. 1990; Askjaer and Kjems 1998). Of these four stem-loops, stem-loop D is primarily responsible for the nuclear export activity of RxRE (Toyoshima et al. 1990; Gröne et al. 1994). However, the other three stem-loops must be present so that RxRE RNA folds into the form of the four stem-loop structure, which is optimal for nuclear export. One hypothesis is that this structure stabilizes stem-loop D and/or interacts with unidentified nuclear factors that facilitate Rex-dependent mRNA transport.

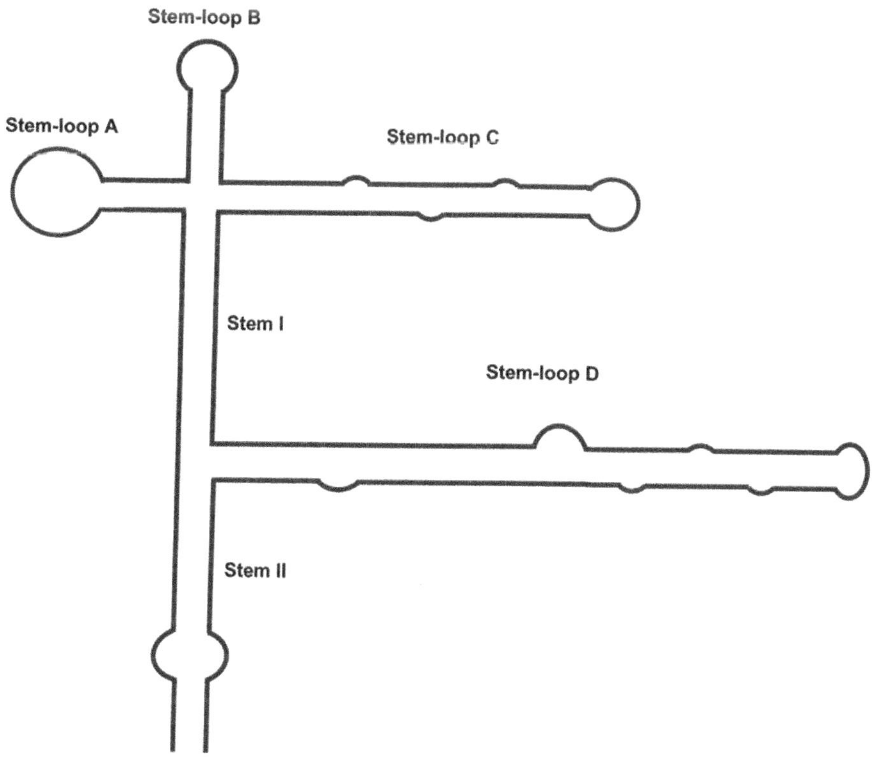

Figure 3.16. *RxRE RNA folding. Diagram of RxRE folding consisting of stem-loops A–D and stems I and II*

The formation of the Rex–RxRE–Crm1 complex has been much less studied than the equivalent HIV-1 complex. Therefore, there is no model describing in detail the steps leading to the formation of the Rex–RxRE–Crm1 complex. However, studies of the Rex–RxRE interaction have shown that Rex binds only to stem-loop D (Bogerd et al. 1991; Askjaer and Kjems 1998). Rex protein can bind to the isolated stem-loop D but its affinity for it is greatly diminished compared to the affinity it has for the entire RxRE region (Askjaer and Kjems 1998). The four-loop stem folding of RxRE thus plays an important role in its interaction with Rex. Two studies (Bogerd et al. 1991; Baskerville et al. 1995) have identified a two-nucleotide bulge (see Figure 3.17, site 1) as the unique site of interaction with Rex. A second high-affinity site (site 2) was characterized by another study that was performed under different experimental conditions than the previous two studies (Askjaer and Kjems 1998). Stem-loop D therefore binds at least two Rex proteins and no more than four through protein–protein interactions a priori. Crm1 could induce Rex oligomerization (Hakata et al. 1998). To date, the process of Crm1–Rex–RxRE complex formation has not been determined.

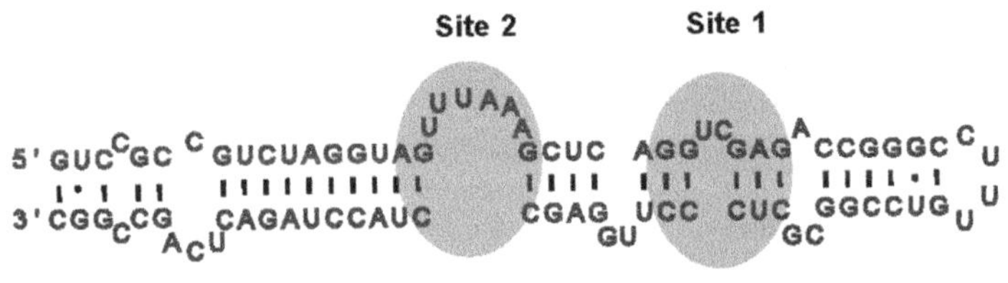

Figure 3.17. *Secondary structure of stem-loop D. This diagram shows the stem-loop D of the RxRE region that plays a crucial role in Rex binding. The two sites with high affinity for Rex are circled in pink. Site 1 has been identified in several studies, while site 2 has been characterized in only one study. For a color version of this figure, see www.iste.co.uk/fosse/structures.zip*

3.3.2.2. *Export of unspliced HIV-1 RNA*

The Rev protein of HIV-1 plays a crucial role in the export of unspliced or partially spliced RNAs from this virus (Malim et al. 1989; Kuzembayeva et al. 2014). Rev is a regulatory protein that is expressed during the early phase of viral replication. Indeed, it is translated from spliced viral mRNAs of about 2 kb that, like those encoding the Tat protein, are the first to be transported to the cytoplasm after transcription of the proviral DNA. These spliced viral mRNAs use the same export pathway to the cytoplasm as that used by cellular mRNAs.

Figure 3.18. *Structural organization of Rev. The half of the protein that is structured contains the bipartite oligomerization domain and the arginine-rich domain that includes the NLS. The other half that is disordered contains the NES. The domains are not shown to scale. For a color version of this figure, see www.iste.co.uk/fosse/structures.zip*

Rev is a small protein (116 amino acids) whose N-terminal half is structured, while the C-terminal half is disordered. This protein, despite its small size, has several functional domains (see Figure 3.18). Indeed, Rev contains an NLS that allows it to be imported into the nucleus. This signal is included in an arginine-rich domain, enabling Rev to bind to RNA. Rev also contains a bipartite oligomerization domain and a leucine-rich NES. Through the NLS and NES signals, Rev can shuttle between the cytoplasm and nucleus (Rausch and Le Grice 2015).

Transport from the nucleus to the cytoplasm of unspliced and partially spliced viral RNAs (approximately 4 kb in size) requires the interaction of Rev with the RRE region (351 nucleotides) that is present in these RNAs (Malim et al. 1989; Rausch and Le Grice 2015). RRE can form two secondary structures that are approximately in equimolar amounts (Watts et al. 2009; Rausch and Le Grice 2015; Sherpa et al. 2015). In both structures, the 3' and 5' ends form a long stem that extends into a four or five stem-loop conformation (see Figure 3.19). The two structures differ only in stem-loops III and IV. The five stem-loop conformation is the one required for optimal HIV-1 replication (Sherpa et al. 2015). The RRE regions of SIV and HIV-2 lentivirus also adopt five stem-loop conformations (Lusvarghi et al. 2013; Pollom et al. 2013). The RRE region of HIV-1 has been studied twice using the small-angle X-ray scattering (SAXS) technique, which provides information on the overall shape of the RNA. One of the studies was performed with the truncated RRE region at the lower part of stem I (Fang et al. 2013), while the other was performed with the entire RRE region (Bai et al. 2014). The truncated RRE region adopts a three-dimensional folding that is A-shaped, while the three-dimensional folding of the whole RRE region resembles a modified A (see Figure 3.20). These conformations result from coaxial stacking between the stems of the RRE region.

The binding of Rev to RRE is cooperative and sequential. Rev oligomerizes onto RRE one molecule at a time. In a first step, one molecule of Rev binds first to the site in stem IIB and then forms a dimer with the second Rev molecule which binds at the junction of the IA, IIB and IIC stems. Oligomerization of Rev continues to form an RRE–Rev complex consisting of one RRE molecule and at least six Rev molecules (Daugherty et al. 2010a). The determination of the exact number of Rev by RRE is a scientific controversy, as the result obtained differs depending on the experimental methods used (Rausch and Le Grice 2015).

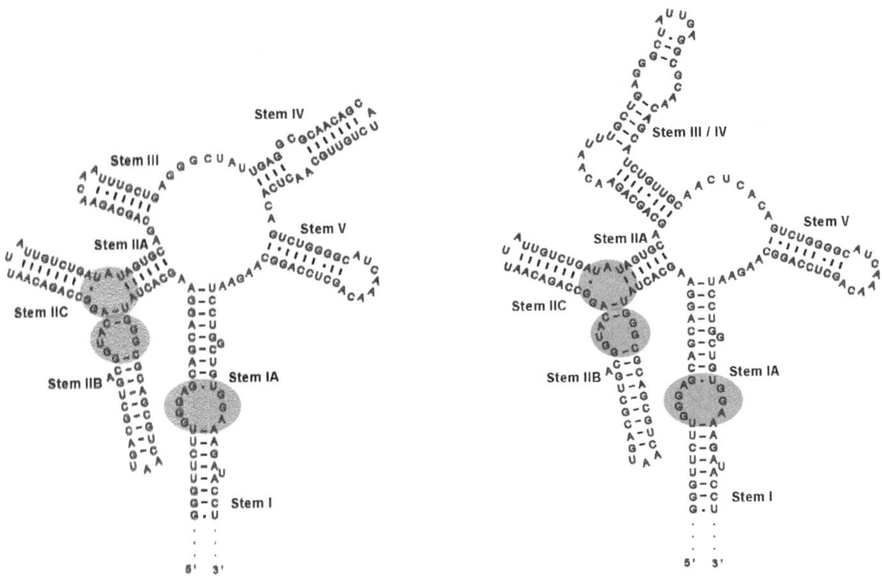

Figure 3.19. *Secondary structures of the RRE. For a color version of this figure, see www.iste.co.uk/fosse/structures.zip*

COMMENTARY ON FIGURE 3.19.– *This diagram shows the part of the RRE region that plays a crucial role in the attachment of Rev. Stem I of the entire RRE region is much longer than that shown here. RRE can form a secondary structure with five stem-loops (left) or four stem-loops (right). The base pairing interactions between stems IIA and V are different in the two conformations. The three sites with high affinity for Rev are circled in pink.*

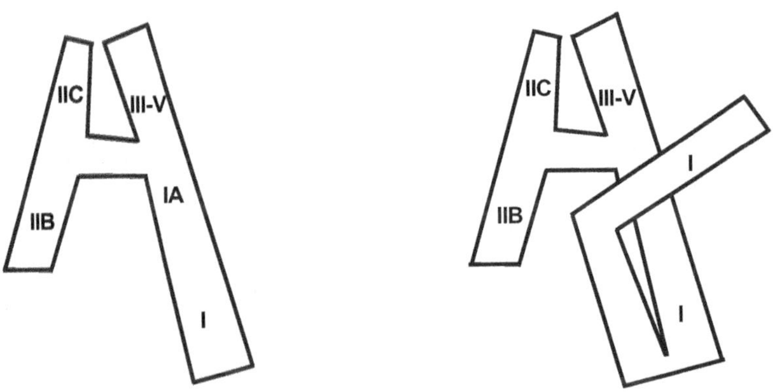

Figure 3.20. *Models of the three-dimensional folding of RRE*

COMMENTARY ON FIGURE 3.20.– *The three-dimensional foldings of the truncated and whole RRE regions are schematized by an A (left) and a modified A (right), respectively. In the whole RRE region, the long stem I, which is much shorter in the truncated RRE region, folds and establishes tertiary interactions with nucleotides forming the stem IA. The stems (I, IA, etc.) constituting the topological elements are indicated within them.*

A model for the formation of the Rev–RRE complex (see Figure 3.21) was proposed based on SAXS data and analysis of Rev binding to the whole RRE RNA using the SHAPE method (Bai et al. 2014). The model is characterized by a change in the three-dimensional folding of RRE induced by the formation of the Rev tetramer. This conformational change makes a site of Rev accessible that is located in stem I.

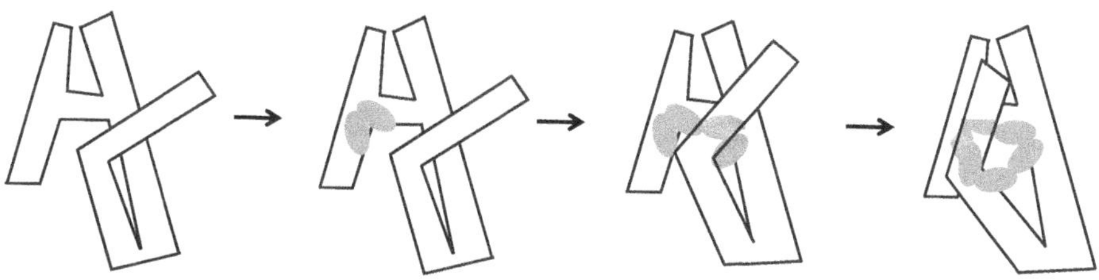

Figure 3.21. *Model of Rev binding to preorganized RRE RNA. The RRE RNA, having a preorganized three-dimensional A-shaped structure, guides the sequential binding of Rev until a hexamer is formed that is competent to bind a Crm1–Ran^{GTP} dimer via two of its six NESs. Each Rev molecule is represented by a yellow oval shape. For a color version of this figure, see www.iste.co.uk/fosse/structures.zip*

To date, the three-dimensional structure of the large and flexible Rev–RRE–Crm1–Ran^{GTP} complex has not been determined. However, a structural model of the Rev–RRE–Crm1–Ran^{GTP} complex has been proposed based on biochemical data and electron microscopy data and through the determination by X-ray crystallography of the structure of the dimer formed by Rev (Daugherty et al. 2010b; Booth et al. 2014).

In this jellyfish-like model, one side of the Rev hexamer interacts with RNA via the arginine-rich region, while the opposite side projects outward the six tentacles consisting of the six disordered C-termini, each containing an NES (see Figure 3.22). The Rev–RRE complex binds a Crm1–Ran^{GTP} dimer via two of its six NESs and can thus pass through the nuclear pore.

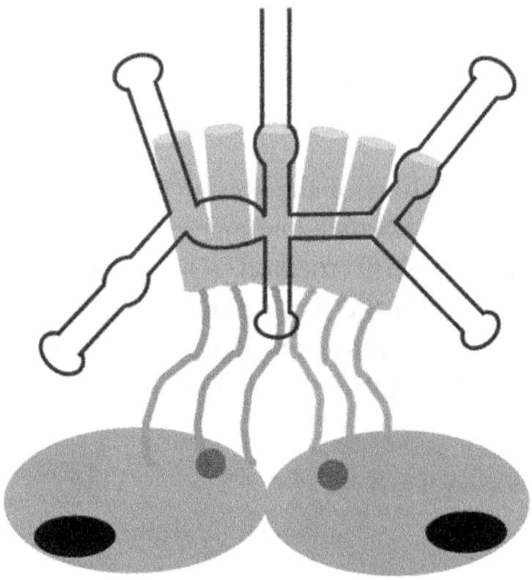

Figure 3.22. *Model of the Rev–RRE–Crm1–Ran^{GTP}complex. For a color version of this figure, see www.iste.co.uk/fosse/structures.zip*

COMMENTARY ON FIGURE 3.22.– *The hexamer of Rev adopting a jellyfish-like structure is colored yellow. The structured part of each subunit of the hexamer is schematized by a cylinder, while the disordered tentacle-like part corresponds to a thick line. The secondary structure of RRE RNA forming the five stem-loop conformation is in blue. Each subunit of the Crm1 dimer is represented by a gray oval shape. The Ran^{GTP} associated with Crm1 corresponds to the black oval shape. The two purple disks indicate the two sites through which the Rev–RRE complex is tied to the Crm1–Ran^{GTP} dimer.*

3.4. RNA structures regulating the translation of retroviral RNAs

In eukaryotic cells, translation is a process that can be divided into four main phases: initiation, elongation, termination and recycling of ribosomes. Translational regulation is mainly carried out at initiation and consists of several steps leading to the assembly of the 80S ribosome at the translation start codon (almost exclusively the AUG triplet). In the vast majority of cases, the initiation of translation is dependent on the 5' cap ($mGppp^7$). This recruits the 40S subunit of the ribosome via its interaction with the translation initiation factor eIF4E. Cap-independent translation initiation has been discovered in picornaviruses, whose mRNAs do not have a cap at their 5' end (Jang et al. 1988; Pelletier and Sonenberg 1988). In picornaviruses, translation initiation depends primarily on a structured region of the mRNA that constitutes an internal ribosome entry site (IRES). Ten to fifteen percent of cellular mRNAs contain an IRES, although they possess a cap (Barrera et al. 2020a). The presence of a cap and an IRES on the same mRNA allows for a switch from cap-dependent to IRES-dependent translation initiation when the cell is under stress. The activity of an IRES requires it to interact with factors called IRES trans-acting factors (ITAFs). ITAFs are usually chaperone proteins that allow an IRES to adopt the functional conformation.

Retroviruses, all of whose mRNAs possess a cap, use the translation machinery of the eukaryotic cells they infect. IRESs have been identified in both simple and complex retroviruses (Barrera et al. 2020a). Translation initiation of retroviral mRNAs has been studied mainly in HIV-1 and HIV-2, which are complex viruses. However, a few studies on translation initiation have been performed with simple retroviruses.

3.4.1. *IRESs of simple retroviruses*

The first retroviral IRESs were identified in simple retroviruses belonging to the genera gammaretrovirus and alpharetrovirus (Berlioz and Darlix 1995; Vagner et al. 1995; López-Lastra et al. 1997; Deffaud and Darlix 2000a, 2000b). They are in the structured regions that are formed by the 5'-UTR domains of retroviral mRNAs. With the exception of the IRES that is present in the F-MLV *env* mRNA, the IRESs which have been characterized are located in the unspliced mRNAs (see Figure 3.23). A common structure for the different IRESs was not identified in the 5'-UTR regions of the simple retroviruses studied.

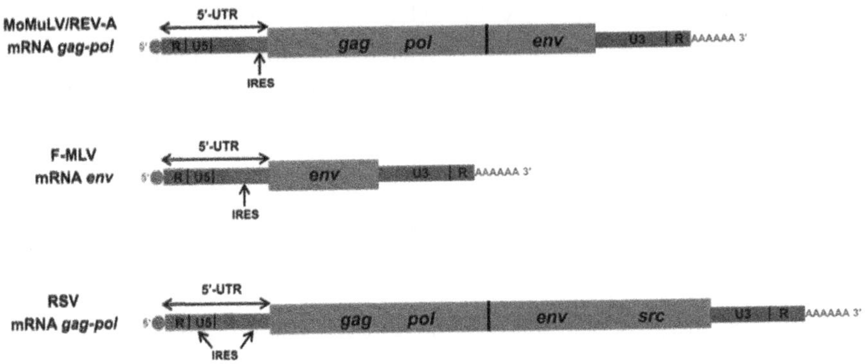

Figure 3.23. *Localization of IRES in single retrovirus mRNAs. For a color version of this figure, see www.iste.co.uk/fosse/structures.zip*

COMMENTARY ON FIGURE 3.23.– *The gag–pol mRNAs of gammaretroviruses (MoMuLV and REV-A) and an alpharetrovirus (RSV) are shown as examples. The env mRNA of F-MLV, a gammaretrovirus, is also shown. IRESs, green regions, are indicated by arrows. Regions and genes are not shown to scale. The cap is indicated by an orange disk.*

3.4.1.1. *IRESs of gammetrovirus*

The 5'-UTR domain of the unspliced mRNAs of two murine retroviruses (MoMuLV and F-MLV) and one avian retrovirus (REV-A) has IRES activity (Berlioz and Darlix 1995; Vagner et al. 1995; López-Lastra et al. 1997). This activity depends mainly on a region just upstream of the *gag* gene initiation codon, which consists of 126 (MoMuLV) or 129 (REV-A) nucleotides (Vagner et al. 1995; López-Lastra et al. 1997). In the case of MoMuLV, part of this region can form two stem-loops, one of which carries a pyrimidine motif (see Figure 3.24) that plays an important role in IRES activity. An interaction between this motif and the splicing factor PTB has been demonstrated (Vagner et al. 1995). Interestingly, the IRES activity of several picornaviruses requires a pyrimidine and PTB motif that is part of the ITAFs (Barrera et al. 2020a). Other ITAFs involved in the IRES activity of unspliced gammaretrovirus mRNAs have not been identified. The study that identified an IRES in the 5'-UTR domain of F-MLV *env* mRNA is the only one to date that shows that the initiation of translation of a gammaretrovirus mRNA can depend on an IRES (Deffaud and Darlix 2000a). This IRES, whose secondary structure has not been determined, is located between the splice donor/acceptor site and the initiation codon of the *env* gene (see Figure 3.23). The possibility that the mRNAs of many simple retroviruses possess an IRES cannot be ruled out, because the IRES activity of these mRNAs has hardly been investigated.

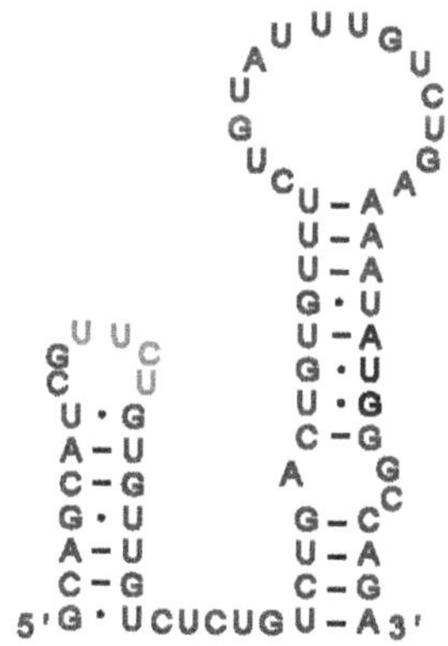

Figure 3.24. *Secondary structure of the 3' end of the IRES of the MoMuLV unspliced mRNA. The 3' end of the IRES of MoMuLV unspliced mRNA can form two stem-loops (Mougel et al. 1993). The pyrimidine motif interacting with PTB is in orange. The initiation codon of the* gag *and* gag–pol *genes is in black. For a color version of this figure, see www.iste.co.uk/fosse/structures.zip*

3.4.1.2. *IRESs of an alpharetrovirus*

The 5'-UTR region of the unspliced mRNA of RSV, an alpharetrovirus, contains an IRES that is bipartite (see Figure 3.23). Both parts, when isolated, each exhibit IRES activity, which is, however, weaker for the 5' part than for the 3' part (Deffaud and Darlix 2000b). The secondary structures of the 5'-UTR region and the ITAFs that are required for IRES activity have not been characterized.

3.4.2. *Translation initiation in complex retroviruses*

Apart from lentiviruses, and more particularly HIV-1, few studies have been devoted to the identification of mechanisms regulating the initiation of translation of complex retroviruses mRNAs. With the exception of HIV-2, all of the complex retroviruses studied have an IRES in the 5'-UTR domain of the unspliced mRNA (see Figure 3.25).

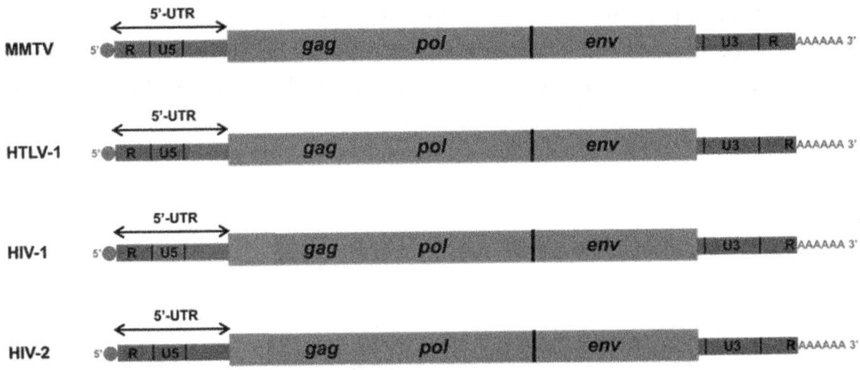

Figure 3.25. *Localization of IRESs in unspliced mRNAs of complex retroviruses. For a color version of this figure, see www.iste.co.uk/fosse/structures.zip*

COMMENTARY ON FIGURE 3.25.– *The gag–pol mRNAs of MMTV, HTLV-1, HIV-1 and HIV-2 retroviruses are shown as examples. IRESs in the 5'-UTR domains are in green. IRESs in the gag gene are in orange. In the case of HIV-2, the orange part corresponds to three IRESs. Regions and genes are not shown to scale. The cap is indicated by an orange disk.*

3.4.2.1. *IRESs of MMTV*

The 5'-UTR domain (320 nucleotides) of the unspliced mRNA of MMTV has an IRES (Vallejos et al. 2010). It has not been precisely localized, and its secondary structure has not been determined. PTB, which binds to the 5'-UTR region and stimulates IRES activity, has been identified as an ITAF (Cáceres et al. 2016). The hnRNP A1 protein has been characterized as another ITAF, as it is required for IRES activity (Barrera et al. 2020b).

3.4.2.2. *IREs of HTLV-1*

The 5'-UTR domain of the HTLV-1 unspliced mRNA contains a 181-nucleotide long IRES that is just upstream of the *gag* gene (Olivares et al. 2014). Its secondary structure has not been determined. IRES activity is dependent on the ribosomal protein S25 and is stimulated by the hnRNP A1 protein, which is an ITAF (Olivares et al. 2014; Barrera et al. 2020b).

The HBZ regulatory protein is encoded by the negative polarity strand of the HTLV-1 genome. Two isoforms of HBZ are synthesized from two antisense mRNAs, which are either spliced or unspliced. The first 218 nucleotides of the 5'-UTR region of the spliced antisense mRNA constitute an IRES (Cáceres et al. 2018). The IRES folds as three domains (see Figure 3.26). Domain I plays the most important role in IRES activity. This is dependent on the ribosomal protein S25 and the translation initiation factor eIF5A. Interactions of these or other proteins with IRES have not been reported in the studies published to date.

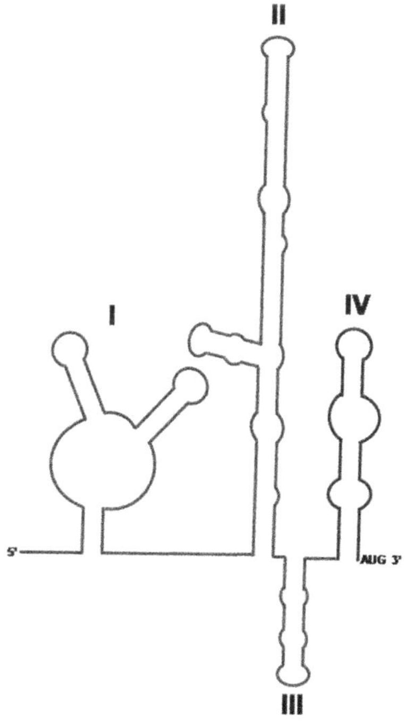

Figure 3.26. *Folding of the 5'-UTR domain of spliced HBZ mRNA. The IRES, shown in green, consists of domains I, II and III. Domain I contains two stem-loops and a large inner loop. Domain IV, in blue, corresponds to the 3' end of the 5'-UTR that is just upstream of the AUG codon of the HBZ gene. For a color version of this figure, see www.iste.co.uk/fosse/structures.zip*

3.4.2.3. *IRESs of HIV-2*

To date, there are no studies that have focused on the initiation of translation of HIV-2-spliced mRNAs. HIV-2 differs in its mode of initiation of unspliced mRNA translation from other simple and complex retroviruses that have been studied. Indeed, IRES activity in the 5'-UTR domain of the HIV-2 unspliced mRNA could not be demonstrated (Barrera et al. 2020a).

Furthermore, translation initiation does not appear to be dependent on the presence of the 5'-cap at any point in the replicative cycle of the virus.

The initiation of translation of the unspliced HIV-2 mRNA depends primarily on the three IRESs that are located in the 5' end of the coding portion of the *gag* gene (Herbreteau et al. 2005; Ricci et al. 2008; Weill et al. 2010). The region containing these IRESs is approximately 350 nucleotides long and folds into six main domains (see Figure 3.27). The first IRES is downstream of the AUG_1 initiation codon where synthesis of the Gag and Gag–Pol polypeptides begins. The initiation of translation of these polypeptides is original, as it depends on an IRES that is downstream and not upstream of the AUG codon. The AUG_2 and AUG_3 initiation codons allow the synthesis of truncated Gag polypeptides whose roles in virus replication are not known. The second IRES is located between AUG_1 and AUG_2. The third IRES is between AUG_2 and AUG_3 (Ricci et al. 2008).

The three IRESs function independently and are able to recruit three ribosomal translation initiation complexes to the same unspliced mRNA molecule (Weill et al. 2010). The region containing domains P2–P5 (see Figure 3.27) is capable of directly binding the 40S subunit of the ribosome and the translation initiation factor eIF3 (Locker et al. 2011). The interaction sites have not been precisely identified. These results led to the proposal of a two-step sequential process that occurs three times to allow three initiations on the same RNA molecule:

– recruitment by the P2–P5 region of the 40S subunit and the eIF3 factor;

– transfer of the 40S subunit and the eIF3 factor to one of the three AUG codons.

The molecular basis of the mechanism involved in the transfer has not been determined. The transfer may require the disruption of intramolecular base pairing interactions within IRESs because the synthesis of whole and truncated Gag polypeptides requires the presence of eIF4A, which is an RNA helicase (Locker et al. 2011).

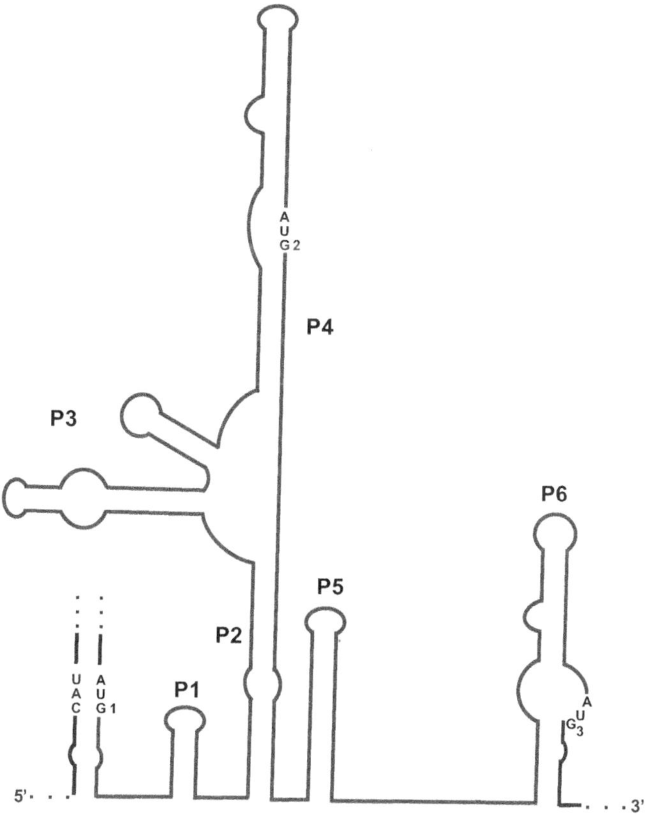

Figure 3.27. *Folding of the 5' end of the HIV-2 gag gene. For a color version of this figure, see www.iste.co.uk/fosse/structures.zip*

COMMENTARY ON FIGURE 3.27.– *The synthesis of Gag and Gag–Pol polypeptides starts at codon AUG1. The synthesis of truncated Gag polypeptides is initiated at codons AUG2 and AUG3. The three IRESs are in the highly structured region outlined in green. The 5'-UTR region is upstream of the AUG1 codon. The sequence downstream of codon AUG3 corresponds to the 3' part of the gag gene.*

3.4.2.4. *IRESs of HIV-1*

Three different mechanisms may be responsible for the initiation of translation of unspliced HIV-1 mRNA:

– canonical initiation dependent on the 5' cap and its direct interaction with eIF4E;

– non-canonical initiation dependent on the 5' cap but not requiring interaction of the latter with the eIF4E factor;

– non-canonical initiation via IRESs (Barrera et al. 2020a).

Studies have shown that the initiation of translation is a dynamic process that is primarily cap-dependent at the onset of HIV-1 infection of the cell (Gendron et al. 2011; Vallejos et al. 2011; Monette et al. 2013). Over time, infection induces a cellular environment (e.g. oxidative stress) that is unfavorable for cap-dependent initiation but allows initiation via IRES.

The TAR stem-loop, which is located at the 5' end of unspliced and spliced mRNAs, is conserved in different HIV-1 isolates because it plays a crucial role in proviral DNA transcription (see section 3.1.2). However, it inhibits translation in vitro because, due to its high stability, it is a barrier during cap recruitment of eIF4E and other factors forming the translation preinitiation complex with the 40S subunit of the ribosome (de Breyne and Ohlmann 2018). The virus can use ex vivo cellular proteins (Staufen-1, UPF1, TRBP, etc.) and viral proteins (Rev and Tat) to overcome the inhibition induced by the TAR structure. These proteins' modes of action have not all been characterized. However, it is known that several of these proteins bind to the TAR and that some possess RNA helicase activity (de Breyne and Ohlmann 2018; Barrera et al. 2020a). They thus act by destabilizing the TAR structure and therefore allowing the translation preinitiation complex to form at the cap and slide along the mRNA.

The initiation of unspliced mRNA translation becomes cap-independent and dependent on two IRESs, which are located in the 5'-UTR domain and the *gag* gene, respectively (see Figure 3.25), when the cellular environment changes and the G2/M transition of the cell cycle is blocked by the HIV-1 Vpr protein (Brasey et al. 2003; Barrera et al. 2020a). The 5'-IRES (approximately 225 nucleotides), which is present in the 5'-UTR region and just upstream of the *gag* gene initiation codon, enables the synthesis of Gag and Gag–Pol polypeptides (Brasey et al. 2003; Gendron et al. 2011; Vallejos et al. 2011). It consists of the four structured domains PBS, DIS, SD and Psi (see Figure 3.28). It is probably the topological arrangement of the four domains that confers IRES activity (Gendron et al. 2011; Vallejos et al.

2011; Plank et al. 2013). The 5'-IRES requires cellular proteins to be functional. Among these, the hnRNP A1 protein was the first ITAF to be characterized (Monette et al. 2009). This protein's mechanism of action, which stimulates the 5'-IRES activity, has not been determined. The cellular protein HuR is an ITAF that inhibits the 5'-IRES activity (Rivas-Aravena et al. 2009). Other cellular proteins (DDX3 helicase and hRIP) have been characterized as ITAFs (Barrera et al. 2020a). The molecular mechanisms that are used by ITAFs to regulate the 5'-IRES activity are not known. The ribosomal protein S25 is involved in the recruitment of the translation preinitiation complex by the 5'-IRES (Carvajal et al. 2016). A direct interaction between this protein and the 5'-IRES probably does not occur as it could not be demonstrated.

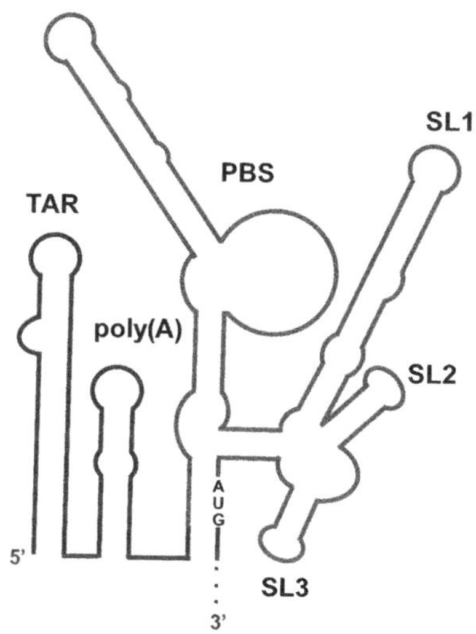

Figure 3.28. *Folding of the 5'-UTR domain of HIV-1 unspliced mRNA. For a color version of this figure, see www.iste.co.uk/fosse/structures.zip*

COMMENTARY ON FIGURE 3.28 – *The 5'-UTR region of the unspliced mRNA can fold into six stem-loops (Wilkinson et al. 2008) of which the two closest to the 5' end are not required for IRES activity. The IRES in green is upstream of the gag gene initiation codon (AUG in black). The apical loop of SL2 contains the splice donor site (SD).*

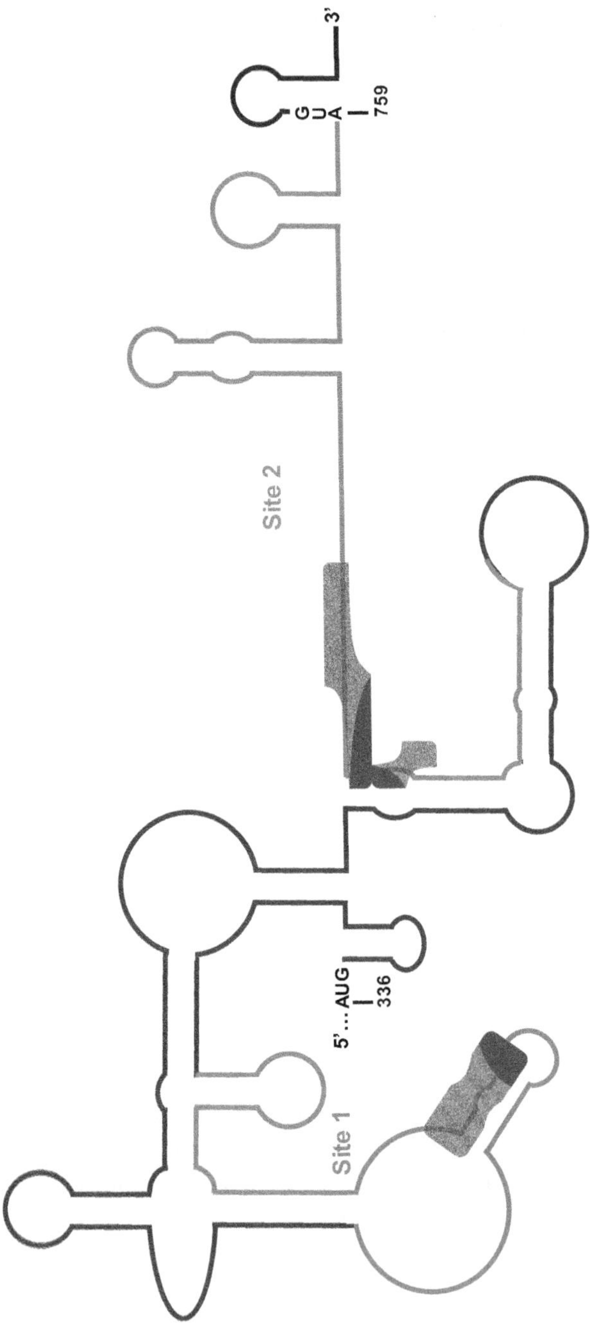

Figure 3.29. *HIV-1 IRES–Gag folding. For a color version of this figure, see www.iste.co.uk/fosse/structures.zip*

COMMENTARY ON FIGURE 3.29.– *The IRES–Gag is located between codons AUG 336 and 759. Codon AUG 336 allows the synthesis of the Gag polypeptide, while codon AUG 759 allows the synthesis of the Gag–p40 protein. Sites 1 and 2 shown in orange correspond to the two sites where the 40S subunit of the ribosome binds. These sites contain an adenine-rich region, which is boxed in pink.*

The IRES–Gag (approximately 420 nucleotides), which is located in the *gag* gene (see Figure 3.25), synthesizes the Gag polypeptide and a truncated form of the Gag polypeptide, which is called Gag–p40 (Buck et al. 2001). The role of the Gag–p40 protein is not known, but it is known that virus infectivity decreases if the protein is not produced. As in HIV-2, the highly structured IRES–Gag (see Figure 3.29) is located downstream of the *gag* gene initiation codon. Two sites binding the 40S subunit of the ribosome have been identified in the IRES–Gag (Deforges et al. 2017). Each site is thought to recruit a 40S subunit of the ribosome via an adenine-rich sequence. Gag–p40 synthesis results from preinitiation complex sliding from site 2 to AUG 759 in the canonical 5'–3' direction. In contrast, a preinitiation complex shift occurs from site 1 to AUG 336 in the non-canonical 3'–5' direction to enable the synthesis of the Gag polypeptide (Deforges et al. 2017). ITAFs involved in IRES–Gag activity have not been identified. To date, we do not know how the 5'–IRES and IRES–Gag coordinately regulate Gag polypeptide synthesis from the same AUG codon.

Alternative splicing in HIV-1 produces the *nef*, *tat*, *vif*, *vpr* and *vpu* mRNAs, which also have an IRES in their 5'-UTR domain (Charnay et al. 2009; Plank et al. 2014). All of these mRNAs share in common with the unspliced mRNA the same sequence (289 nucleotides), which is upstream of the splice donor site (SD) and constitutes part of the 5'-UTR domain. The folding of this sequence, which is the same in the different mRNAs, leads to the formation of the TAR, poly(A), PBS and DIS (Plank et al. 2014). IRES activity does not require the TAR and poly(A) stem-loops but does require the PBS and DIS structures. IRES activity is modulated by the sequence that is downstream of the splice acceptor site and corresponds to the other part of the 5'-UTR domain. ITAFs that are associated with the 5'-IRES of mRNAs have not been characterized.

The 5' part of exon 1 of *tat* has a short sequence (16 nucleotides), whose stem-loop structure called TIM-TAM is highly conserved among different HIV-1 isolates (Khoury et al. 2020). This stem-loop corresponds to an independent structural domain in the gRNA (see Figure 3.30), while it constitutes the apical part of a long stem-loop in the *tat* mRNA. It plays a dual role (Khoury et al. 2020):

– it is necessary for the IRES activity which is present in the 5' coding end of the *tat* gene;

– it blocks the progression of the ribosome when the initiation of the translation of *tat* mRNA depends on the cap.

The proteins involved in IRES activity have not been identified. It has been proposed that in the early phase of infection prior to viral DNA integration, initial Tat protein synthesis is dependent on TIM-TAM, which is present in the gRNA (Khoury et al. 2020).

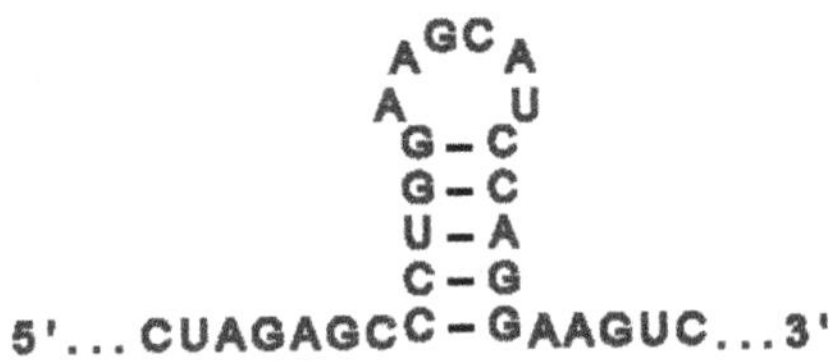

Figure 3.30. *Secondary structure of TIM-TAM in gRNA. The stem-loop TIM-TAM is flanked by two short, unpaired sequences in the gRNA (Watts et al. 2009)*

4

Encapsidation of Genomic RNA in the Retroviral Particle

Studies performed with alpharetroviruses, gammaretroviruses, lentiviruses and various techniques (analytical ultracentrifugation, electron microscopy and agarose gel electrophoresis) showed that the gRNA extracted from the virions is in the form of a dimer whose subunits are linked by non-covalent bonds (Mangel et al. 1974; Bender and Davidson 1976; Bender et al. 1978; Murti et al. 1981; Fu et al. 1994; Höglund et al. 1997). The virion of all retroviral species thus contains a diploid genome that consists of two gRNA molecules. The reverse transcription of the diploid genome generates genetic diversity that allows retroviruses to develop resistance to antiretrovirals and to evade the host immune system (see Chapter 2, section 2.4). With the exception of spumaretroviruses, encapsidation is a process common to all retroviruses that selects and packages into a viral particle two gRNA molecules from the majority of viral and cellular mRNAs. This process requires the binding of the Gag precursor, via its NC domain, to the gRNA. In several retroviral species, the gRNA must first dimerize in order to be recognized by the Gag polypeptide.

4.1. RNA structures and mechanisms governing gRNA dimerization

The gRNA dimer extracted from virions (mature virus particles) is described as stable because dissociation of the subunits requires a relatively high temperature. In contrast, the gRNA dimer extracted from non-infectious immature viral particles is considered unstable because dissociation of the subunits occurs at a significantly lower temperature (Stoltzfus and Snyder 1975; Fu and Rein 1993; Fu et al. 1994). Conversion of the unstable dimer to the stable dimer requires cleavage of the Gag polypeptide by the viral protease (Stewart et al. 1990; Fu and Rein 1993; Fu et al. 1994; Shehu-Xhilaga et al. 2001b). Studies based on site-directed mutagenesis suggest that NC, a cleavage product of the Gag precursor, plays an important role in stabilizing the gRNA dimer (Méric and Spahr 1986; Méric et al. 1988; Fu and Rein 1993).

Electron microscopy has enabled the identification of the main interaction between the subunits of the gRNA dimer that is extracted from the virion (Bender and Davidson 1976; Murti et al. 1981; Höglund et al. 1997). This interaction, termed DLS, occurs near the 5' end of the gRNA in the region containing the encapsidation signal. Since the electron microscopy studies were performed under semi-denaturing conditions, we cannot completely exclude the possibility that these conditions induce conformational changes in the gRNA, allowing it to form a DLS that does not exist in the virion. Dimerization of gRNA from several retroviruses has been reproduced in vitro using short RNAs that were synthesized in vitro and corresponded to the 5' ends of retroviral genomes (Bieth et al. 1990; Darlix et al. 1990; Prats et al. 1990; Roy et al. 1990; Marquet et al. 1991). Numerous in vitro studies have characterized the sequences, structures and mechanisms that regulate gRNA dimerization in several retroviruses.

4.1.1. *Structures and mechanisms in alpharetroviruses*

The gRNA dimer of RSV type Pr-A, an alpharetrovirus, was extracted from the virion and analyzed with electron microscopy (Murti et al. 1981). This analysis indicates that the DLS that associates the two subunits is located approximately 479–539 nucleotides from the 5' end (see Figure 4.1). Dimerization of alpharetrovirus gRNA was studied in vitro using RNAs containing the DLS. In vitro experiments performed under physiological conditions (temperature and salinity), and with RNAs corresponding exactly to the 5' part of alpharetrovirus (ALV and RSV) gRNA, could not reveal a dimerization site at the DLS (Polge et al. 2000; Liu et al. 2020). In contrast, a contact point associating the two subunits of the RNA dimer generated in vitro was identified in the L3 element of ALV (SR-A type) that is approximately 200 nucleotides upstream of the DLS (Fossé et al. 1996; Polge et al. 2000). The L3 element adopts a stem-loop structure (see Figure 4.2) that is conserved among alpharetroviruses. A single-base mutation in the apical loop autocomplementary sequence is sufficient to suppress dimerization in vitro (Polge et al. 2000). Deletion of the L3 element decreases replication of SR-A RSV by 20- to 300-fold (Doria-Rose and Vogt 1998). In addition, there is selection pressure to restore an autocomplementary sequence in the apical loop after several replication cycles when it has been mutated (Doria-Rose and Vogt 1998). The importance of the L3 element in the replication of SR-A-type alpharetroviruses is probably due to its role in gRNA dimerization. However, the very low level of replication in the absence of the L3 element suggests the existence of one or more secondary dimerization sites. In vitro dimerization of an RNA corresponding to the 5' part of the Pr-C RSV genome is not solely dependent on the L3 element. Indeed, the SLA and L3 elements must be simultaneously deleted to suppress in vitro dimerization (Liu et al. 2020). Although the SLA element also corresponds to a stem-loop whose apical loop contains a self-complementing sequence, it has not been shown to be a contact point between the two subunits of the RNA dimer. Furthermore, the infectivity of RSV (Pr-C type) is not diminished by two deletions that delete the SLA and L3 elements (Liu et al. 2020). Thus, it appears that the L3 stem-loop is not equally important in different alpharetrovirus types.

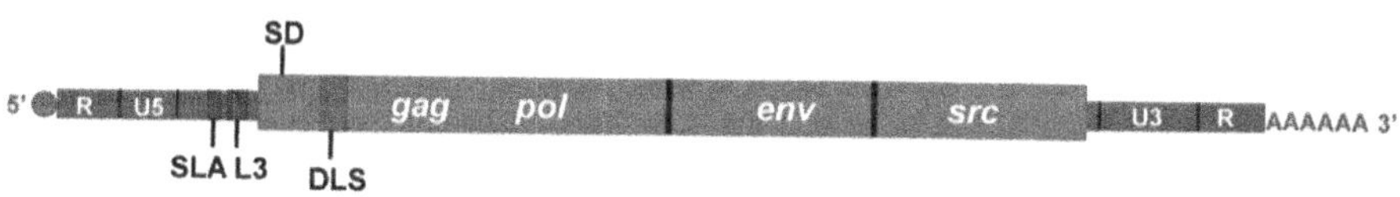

Figure 4.1. *Sites involved in RSV gRNA dimerization. For a color version of this figure, see www.iste.co.uk/fosse/structures.zip*

COMMENTARY ON FIGURE 4.1.– *The DLS identified by electron microscopy is in the gag gene downstream of the splice donor site (SD). The SLA and L3 elements, which are upstream of SD, were identified in vitro using an in vitro synthesized RNA that corresponds to the 5' end of the RSV gRNA. The orange disk represents the 5' cap. Regions and genes are not shown to scale.*

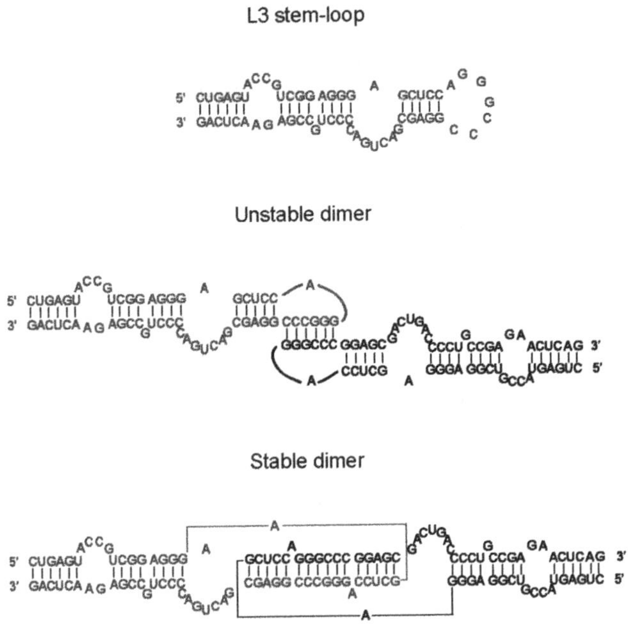

Figure 4.2. *Dimerization of alpharetrovirus gRNA via the L3 element. The secondary structure of the L3 stem-loop in the gRNA monomer is shown in the upper part of the figure. Shown in blue and black are the two L3 stem-loops that link the stable and unstable dimer subunits via Watson–Crick base pairs. For a color version of this figure, see www.iste.co.uk/fosse/structures.zip*

Under physiological temperature and salinity conditions, and in the absence of NC, an RNA, corresponding to the 5' part of ALV gRNA (SR-A type), forms an unstable dimer in vitro via the L3 element (Polge et al. 2000). This dimer may correspond in part to the gRNA dimer present in the immature virus particle. The subunits of this dimer dissociate when analyzed by agarose gel electrophoresis at room temperature and in the absence of magnesium. The apical loop of the L3 stem-loop is directly involved in the loop–loop interaction that links the subunits of the unstable dimer (Polge et al. 2000). This loop–loop interaction, forming a short duplex of six base pairs (see Figure 4.2), is thought to initiate the formation of a long duplex linking the subunits of the stable dimer that is present in the mature virus particle. It is possible to generate a stable dimer using an elevated temperature (60°C) or physiological temperature in the presence of NC (Polge et al. 2000; Ben Ali et al. 2007). The subunits of the stable dimer are linked by a duplex of 16 base pairs that results from the chaperone activity of the NC. Indeed, it opens the apical part of the L3 stem-loop and promotes the formation of a long stable duplex that contains more paired nucleotides than the same sequence in the monomer (see Figure 4.2). Although this duplex could also be generated by the NC in the virion, this has so far neither been demonstrated nor refuted.

4.1.2. *Structures and mechanisms in betaretroviruses*

4.1.2.1. *Dimerization of MPMV gRNA*

The gRNA dimer of MPMV, which is present in the virion, has not been studied by electron microscopy. By analogy with other retroviral species, it is assumed that the 5' part of the genome is responsible for its dimerization. In favor of this hypothesis, an in vitro synthesized RNA corresponding to the 5' part of the MPMV genome can form a dimer in vitro (Aktar et al. 2013). The Pal SL element was identified as the contact point associating the two subunits of the in vitro generated RNA dimer (see Figure 4.3). Although a 35-nucleotide deletion encompassing the Pal SL element decreases MPMV replication by ninefold, this has not been shown to be due to altered gRNA dimerization (Jaballah et al. 2010).

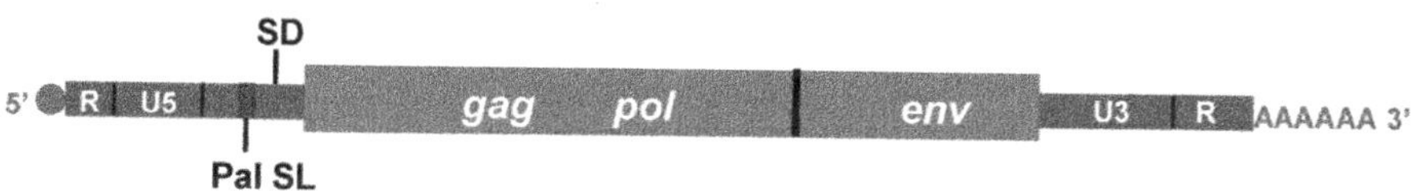

Figure 4.3. *Site involved in the in vitro dimerization of MPMV gRNA. The Pal SL element upstream of the splice donor site (SD) was identified in vitro using an in vitro synthesized RNA that corresponds to the 5' end of MPMV gRNA. The orange disk represents the 5' cap. Regions and genes are not shown to scale. For a color version of this figure, see www.iste.co.uk/fosse/structures.zip*

The Pal SL element forms a short stem-loop that has a self-complementary sequence in the apical loop (see Figure 4.4). A loop–loop interaction via the formation of two base pairs triggers the formation of a duplex of 14 base pairs that links the subunits of the RNA dimer generated in vitro (Aktar et al. 2013).

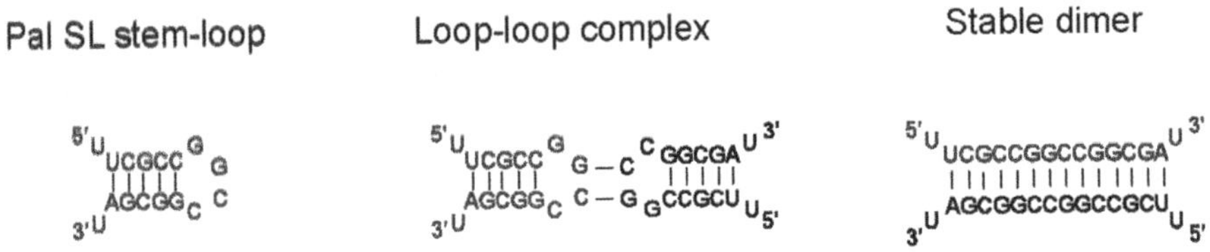

Figure 4.4. *Dimerization of MPMV gRNA via the Pal SL element. For a color version of this figure, see www.iste.co.uk/fosse/structures.zip*

COMMENTARY ON FIGURE 4.4.– *The secondary structure of the Pal SL stem-loop in the gRNA monomer is shown on the left side of the figure. In blue and black are represented the two Pal SL stem-loops that form a loop–loop complex, which then leads to the formation of the stable dimer.*

4.1.2.2. *Dimerization of MMTV gRNA*

The gRNA dimer of MMTV, which is present in the virion, has not been studied by electron microscopy. By analogy with other retroviral species, it is assumed that the 5' part of the genome is responsible for its dimerization. In favor of this hypothesis, an in vitro synthesized RNA corresponding to the 5' part of the MMTV genome can form a dimer in vitro (Aktar et al. 2014). The Pal II element was identified as playing an important role in the formation of the RNA dimer generated in vitro (see Figure 4.5). RNA dimerization is decreased if the PBS Pal element is deleted. The action of this element is rather indirect because it is not a contact point between the two subunits of the dimer. Although deletion of the Pal II element suppresses MMTV replication, this has not been shown to be due to impaired gRNA dimerization (Aktar et al. 2014).

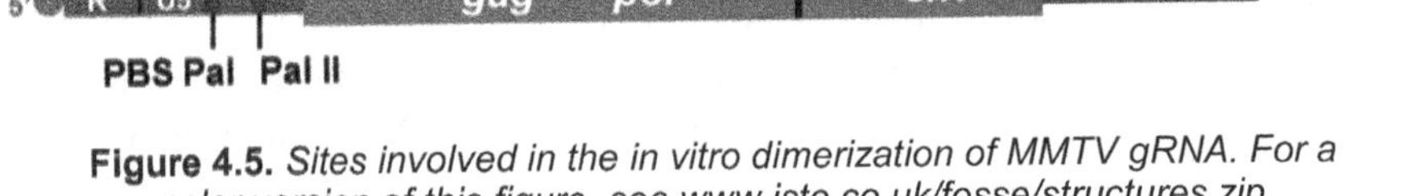

Figure 4.5. *Sites involved in the in vitro dimerization of MMTV gRNA. For a color version of this figure, see www.iste.co.uk/fosse/structures.zip*

COMMENTARY ON FIGURE 4.5.– *The PBS Pal and Pal II elements, which are upstream of the splice donor site (SD), were identified in vitro using an in vitro synthesized RNA that corresponds to the 5′ end of the MMTV gRNA. The orange disk represents the 5′ cap. Regions and genes are not shown to scale.*

The Pal II element forms a short stem-loop that has a self-complementary sequence in the apical loop (see Figure 4.6). The apical loop of the stem-loop is probably directly involved in the intermolecular interaction that links the subunits of the unstable dimer generated in vitro. This loop–loop interaction forming a short duplex of six base pairs is thought to initiate the formation of a long duplex (14 base pairs) linking the subunits of the stable dimer that is observed in vitro (Aktar et al. 2014).

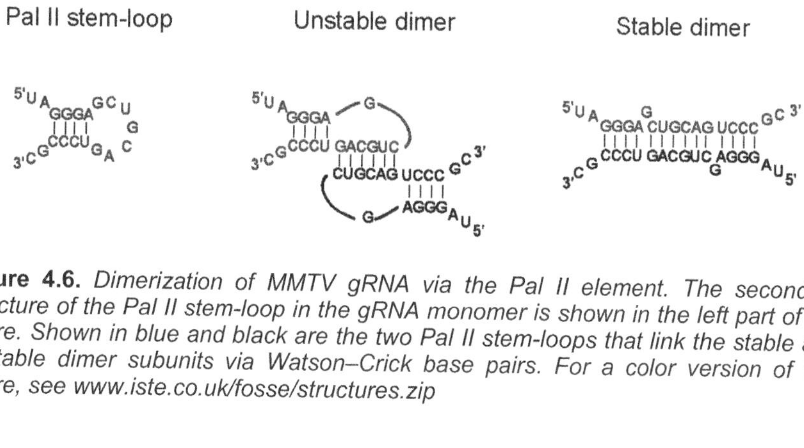

Figure 4.6. *Dimerization of MMTV gRNA via the Pal II element. The secondary structure of the Pal II stem-loop in the gRNA monomer is shown in the left part of the figure. Shown in blue and black are the two Pal II stem-loops that link the stable and unstable dimer subunits via Watson–Crick base pairs. For a color version of this figure, see www.iste.co.uk/fosse/structures.zip*

4.1.3. *Structures and mechanisms in deltaretroviruses*

Dimerization of deltaretrovirus gRNA has been studied mainly in HTLV-1. The gRNA dimer of HTLV-1 has not been studied by electron microscopy. By analogy with other retroviral species, it is assumed that the 5' part of the genome is responsible for its dimerization. In favor of this hypothesis, an RNA synthesized in vitro and corresponding to the 5' part of the HTLV-1 genome can form a dimer in vitro. This dimerization is suppressed by a 32-nucleotide deletion, which is just upstream of the PBS sequence and contains the sequence called DIS2 (see Figure 4.7) (Greatorex et al. 1996; Monie et al. 2001). Deletion of this sequence decreases HTLV-1 replication by only 20–25% (Le Blanc et al. 2000). Therefore, the DIS2 sequence identified in vitro may not correspond to the sequence that links the subunits of the gRNA dimer in the virus particle. In support of this hypothesis, a recent in vitro study, performed under different experimental conditions with RNA containing the entire 5'-UTR region and the 5' end of the *gag* gene, identified the DIS1 sequence as essential for dimerization (Wu et al. 2018). In this study, in vitro dimerization was suppressed by mutations in DIS1 but not by those in DIS2. However, the DIS1 sequence has not been shown to be required for HTLV-1 replication nor to link subunits of the RNA dimer generated in vitro or those of the gRNA dimer that is in the virion.

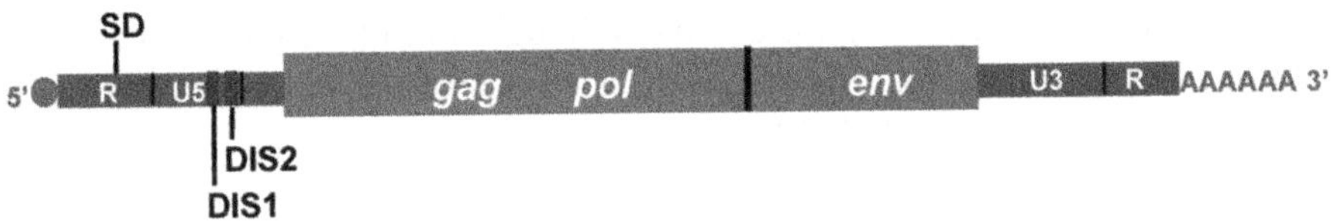

Figure 4.7. *Sites involved in the in vitro dimerization of HTLV-1 gRNA. For a color version of this figure, see www.iste.co.uk/fosse/structures.zip*

COMMENTARY ON FIGURE 4.7.– *The two DIS elements, which are in the U5 domain downstream of the splice donor site (SD), were identified in vitro using an in vitro synthesized RNA that corresponds to the 5' end of the HTLV-1 gRNA. The orange disk represents the 5' cap. Regions and genes are not shown to scale.*

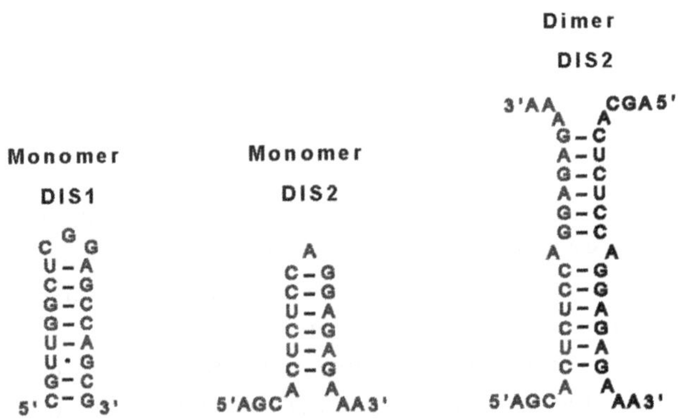

Figure 4.8. *Secondary structures of DIS1 and DIS2 sequences in the gRNA monomer. Shown in blue and black are the two DIS2 sequences that interact to associate the subunits of the dimer generated in vitro. For a color version of this figure, see www.iste.co.uk/fosse/structures.zip*

The DIS1 and DIS2 sequences each form a short stem-loop (see Figure 4.8). DIS2 has an apical loop that consists of a single adenine. This is possible because the CG base pair adjacent to the adenine is unusual. Indeed, guanine adopts the *syn* conformation (Monie et al. 2004). The opening of two DIS2 stem-loops allows the formation of a duplex of 12 base pairs that associates the subunits of the RNA dimer generated in vitro (Monie et al. 2001).

4.1.4. *Structures and mechanisms in gammaretroviruses*

Dimerization and encapsidation of gammaretrovirus gRNA have been studied primarily with MoMuLV. Analysis of gRNA dimers, which were extracted from immature and mature viral particles, suggests that the formation of the stable gRNA dimer in the virion depends on the NC that is released upon cleavage of the Gag precursor by the viral protease (Fu and Rein 1993). Electron microscopy analysis of the MoMuLV gRNA dimer, which was extracted from the virion, indicates that the DLS that associates the two subunits involves less than 50 nucleotides and is located approximately 466 nucleotides from the 5' end (Murti et al. 1981) (see Figure 4.9). MoMuLV gRNA dimerization was studied in vitro using in vitro synthesized RNAs containing the DLS site. In vitro experiments did not reveal a dimerization site at the DLS (Prats et al. 1990; Tounekti et al. 1992; Girard et al. 1995). Instead, four contact points (SLA, SLB, SLC and SLD) associating the two subunits of the RNA dimer generated in vitro were identified upstream of the DLS (Girard et al. 1995; De Tapia et al. 1998; Oroudjev et al. 1999; Miyazaki et al. 2010).

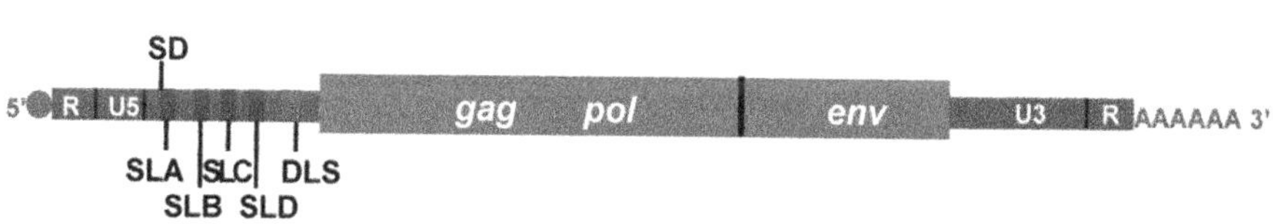

Figure 4.9. *Sites involved in the dimerization of MoMuLV gRNA. For a color version of this figure, see www.iste.co.uk/fosse/structures.zip*

COMMENTARY ON FIGURE 4.9.– *The DLS interaction site identified by electron microscopy is in the encapsidation domain downstream of the splice donor site (SD). The SLA, SLB, SLC and SLD elements, which are also in the encapsidation domain and more than 100 nucleotides away from the DLS, were identified in vitro. The orange disk represents the 5′ cap. Regions and genes are not shown to scale.*

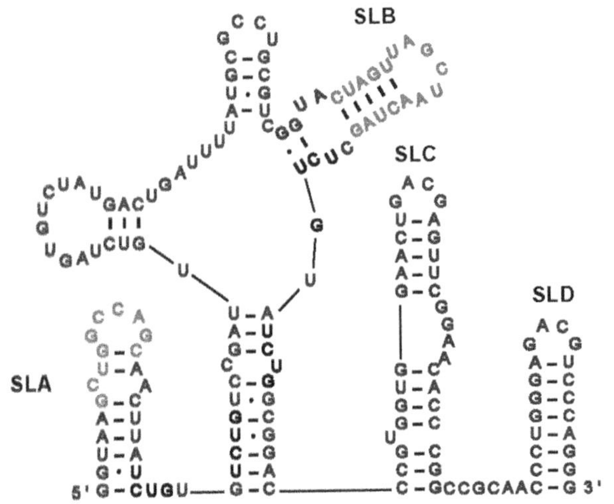

Figure 4.10. *Secondary structure of a part of the MoMuLV gRNA encapsidation domain. For a color version of this figure, see www.iste.co.uk/fosse/structures.zip*

COMMENTARY ON FIGURE 4.10.– *The secondary structure that is shown corresponds to that present in the gRNA monomer. The self-complementary sequences are in orange. The UCUG motif in black, which is repeated four times, is a high-affinity site for the MoMuLV nucleocapsid protein. These motifs are not accessible to the NC in the gRNA monomer because they are engaged in base pairs.*

The SLA, SLB, SLC and SLD elements in the gRNA monomer correspond to stem-loop structures (see Figure 4.10) (Tounekti et al. 1992; Gherghe et al. 2010a). The stem-loops SLA and SLB, which each possess a self-complementary sequence, play a crucial role in in vitro gRNA dimerization (Girard et al. 1995; Oroudjev et al. 1999; Ly and Parslow 2002). The gRNA deleted from both stem-loops is unable to dimerize in vitro and forms a less stable dimer in the virion (Ly and Parslow 2002). Virus replication is greatly diminished by deletion of the SLA stem-loop but not the SLB stem-loop (Aagaard et al. 2004; Grohman et al. 2014). Thus, gRNA dimerization via the SLB stem-loop is not essential for virus viability. The SLC and SLC stem-loops, whose apical loop has the conserved GACG sequence, contribute to the stability of the gRNA dimer by being involved in two loop–loop interactions forming a very short duplex of two base pairs (Miyazaki et al. 2010).

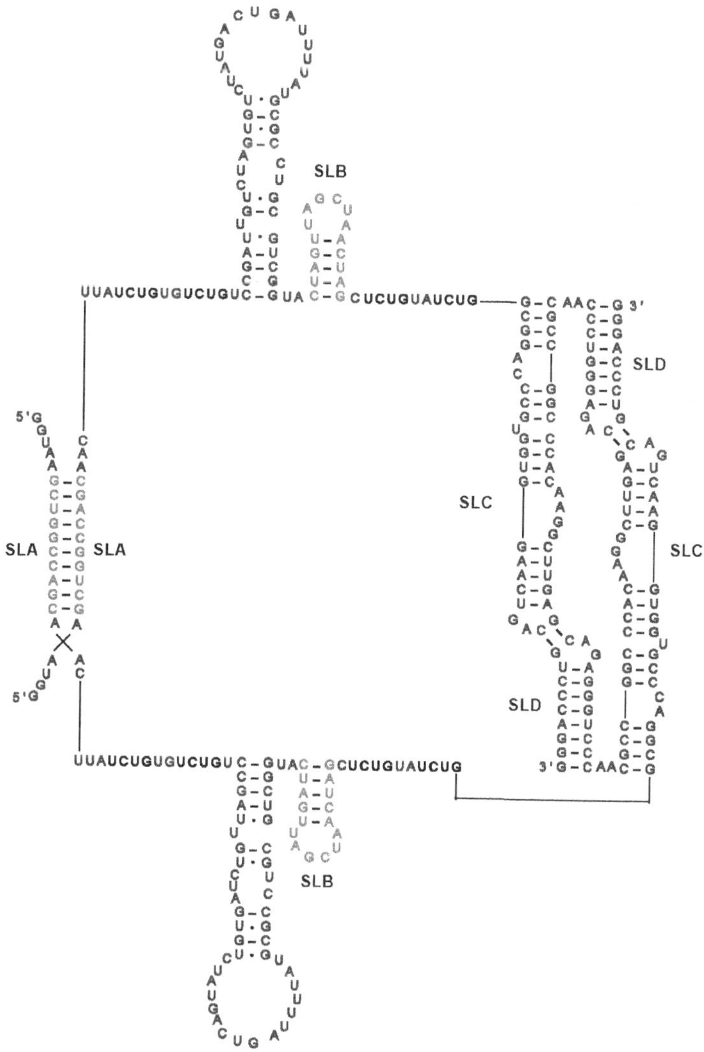

Figure 4.11. *gRNA dimer in the immature virus particle. For a color version of this figure, see www.iste.co.uk/fosse/structures.zip*

COMMENTARY ON FIGURE 4.11.– *The secondary structure that is shown corresponds to that of the gRNA in the immature virus particle. The self-complementary sequences are in orange. The UCUG motif in black that is repeated four times is a high-affinity site for the MoMuLV NC. The NC can interact with all four motifs because they are not involved in base pairs.*

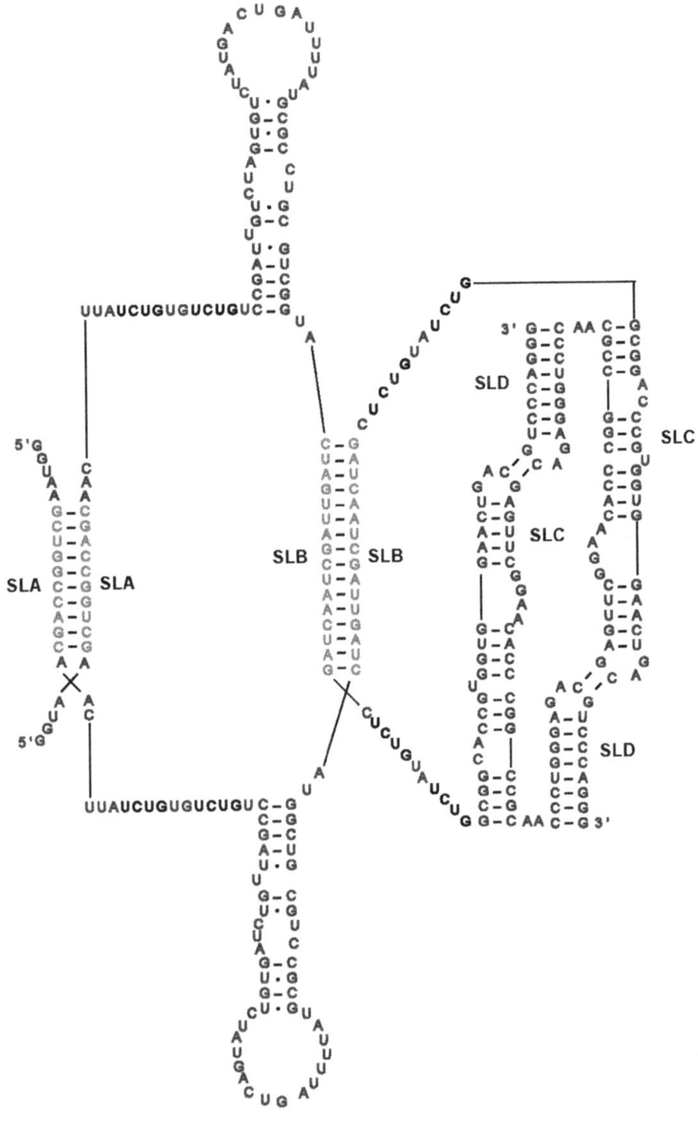

Figure 4.12. *gRNA dimer in the mature virus particle. For a color version of this figure, see www.iste.co.uk/fosse/structures.zip*

COMMENTARY ON FIGURE 4.12.– *The secondary structure that is shown corresponds to that of the gRNA in the mature virus particle. The self-complementary sequences are in orange. The UCUG motif in black that is repeated four times is a high-affinity site for the MoMuLV NC. The NC can interact with all four motifs because they are not involved in base pairs.*

In vitro studies have probably characterized the sequences that link the subunits of the gRNA dimer subunits in the virus particle. Indeed, studies performed using the SHAPE method are in favor of a direct involvement of the SLA and SLB elements in the association of the gRNA subunits that is present in both mature and immature virus particles (Gherghe et al. 2010a, 2010b; Grohman et al. 2014). Specifically, the two subunits of the gRNA dimer, which is present in the immature particle, are primarily associated by the 10-base pair duplex resulting from the opening and pairing of the two SLA stem-loops (see Figure 4.11). The dimer is stabilized by two loop–loop interactions involving the SLC and SLD stem-loops.

The two subunits of the gRNA dimer, which is present in the mature particle, are associated by the duplex of 10 base pairs SLA–SLA and the duplex of 16 base pairs resulting from the opening and pairing of the two SLB stem-loops (see Figure 4.12). NC is likely responsible for the formation of the SLB–SLB duplex (Girard et al. 1996; Grohman et al. 2014). Both duplexes confer high stability to the gRNA dimer. In addition, the dimer is also stabilized by two loop–loop interactions involving the SLC and SLD stem-loops.

4.1.5. *Structures and mechanisms in lentiviruses*

4.1.5.1. *Dimerization of HIV-2 gRNA*

The HIV-2 gRNA dimer has not been studied by electron microscopy. By analogy with other retroviral species, it is assumed that the 5' part of the genome is responsible for its dimerization. In favor of this hypothesis, RNAs synthesized in vitro and containing the 5' part of the HIV-2 genome can form dimers in vitro. The 5' end of the PBS sequence and the TAR3, PAL and SL1 elements (see Figure 4.13) have been identified as playing a role in in vitro dimerization (Dirac et al. 2001; Jossinet et al. 2001; Lanchy et al. 2003;

Purzycka et al. 2011). It has not been determined whether TAR3 and the 5' end of PBS associate the subunits of the gRNA dimer that is in the mature and immature viral particles. Mutations designed to prevent dimerization via the SL1 element have no effect on HIV-2 replication or on the formation of the gRNA dimer that is in the virion (L'Hernault et al. 2007). However, analysis of RNA from heterozygous viral particles suggests that the formation of a six-base pair duplex via two SL1 elements (described below) is involved in the selective encapsidation of dimeric gRNA (Ni et al. 2011). Mutations designed to prevent dimerization via the PAL element inhibit virus replication and significantly decrease the amount of dimeric gRNA in virions (L'Hernault et al. 2007, 2012).

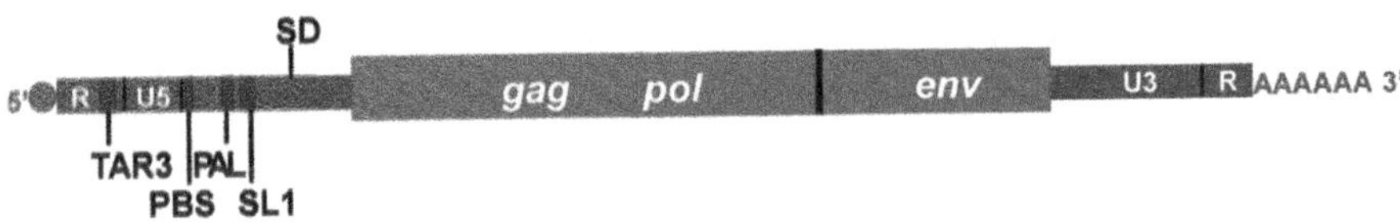

Figure 4.13. *Sites involved in HIV-2 gRNA dimerization. The 5' end of the PBS sequence and the TAR3, PAL and SL1 elements, which are upstream of SD, were identified in vitro. The orange disk represents the 5' cap. Regions and genes are not shown to scale. For a color version of this figure, see www.iste.co.uk/fosse/structures.zip*

At physiological temperature and in the absence of the HIV-2 nucleocapsid protein, an RNA containing the entire 5'-UTR region and the start of the *gag* gene forms an unstable dimer in vitro via TAR3 and the 5' end of PBS and SL1 (Lanchy and Lodmell 2002; Purzycka et al. 2011). These three elements, two of which are stem-loops (see Figure 4.14), associate the subunits of the unstable dimer by forming short duplexes consisting of six to eight base pairs. It has not been determined whether these elements participate in the formation of an unstable gRNA dimer in the immature virus particle.

In vitro, at physiological temperature, the conversion of the unstable dimer into a stable dimer requires the presence of the NC, which, in a first step, destabilizes the intramolecular base pairing interaction constituted by the C-box and G-box sequences (Lanchy et al. 2003; Purzycka et al. 2011). In a second step, the NC, by binding to the upper part of the two SL1 stem-loops, destabilizes the lower part of the latter and thus allows the release of the two PAL sequences that are partially paired (see Figure 4.15). In a third step, the PAL sequences pair to form a duplex of 10 base pairs that are the main contact point of the stable dimer generated in vitro.

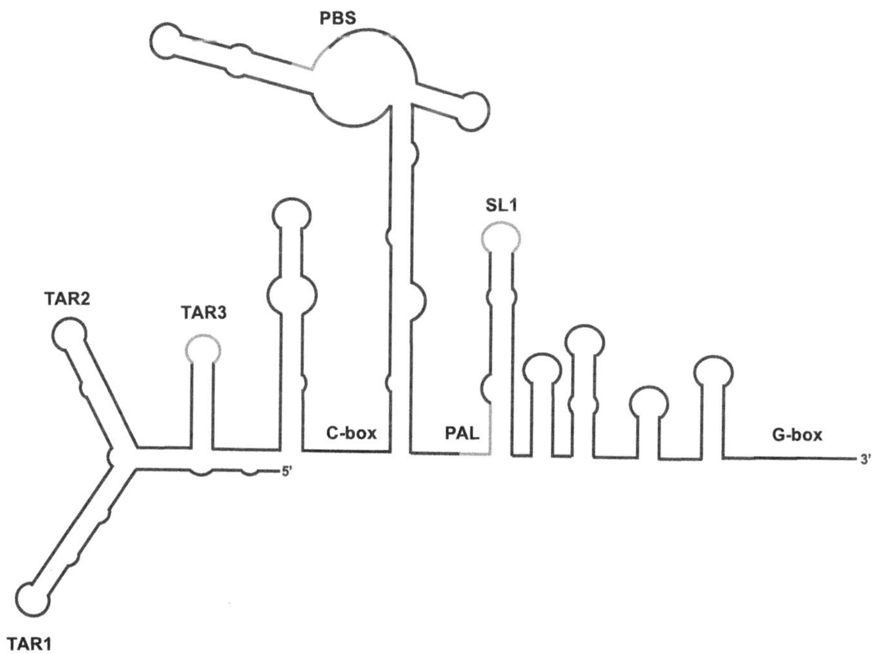

Figure 4.14. *Folding of the 5′ end of HIV-2 gRNA. For a color version of this figure, see www.iste.co.uk/fosse/structures.zip*

COMMENTARY ON FIGURE 4.14.– *The folding that is represented corresponds to that adopted by the monomeric RNA synthesized in vitro and corresponding to the 5′ end of the gRNA. The self-complementary sequences are in orange (10 nucleotides for the PAL sequence, 8 nucleotides for the apical loop of TAR3 and 6 nucleotides for the 5′ end of the PBS sequence and the apical loop of SL1). The C-box and G-box sequences (black lines) form an intramolecular base pairing interaction in the unstable dimer. The G-box sequence contains the initiation codon of the* gag *gene.*

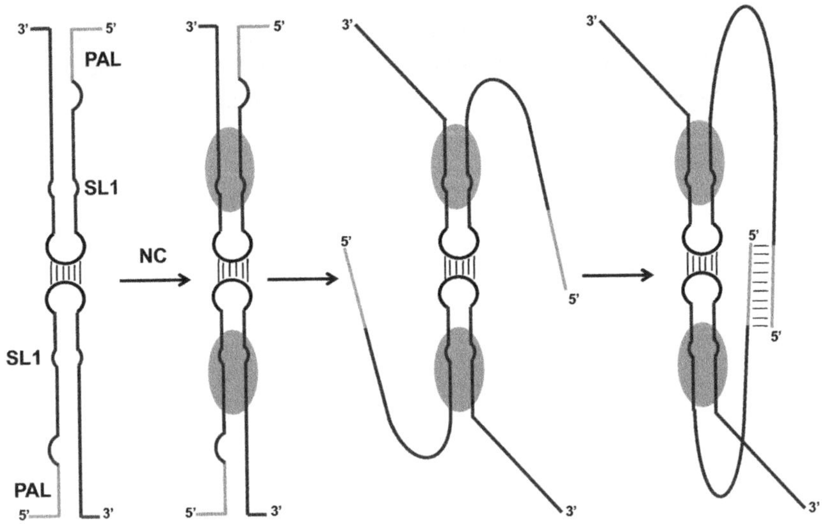

Figure 4.15. *Model for the formation of the stable HIV-2 gRNA dimer via PAL sequence pairing. For a color version of this figure, see www.iste.co.uk/fosse/structures.zip*

COMMENTARY ON FIGURE 4.15.– *In this model, two SL1 stem-loops interact via a loop–loop interaction. This interaction forms a short duplex of six base pairs, which is one of the contact points associating the subunits of the unstable dimer. Apical loops with a self-complementary sequence are represented by a black line. The NC (purple ellipse) binds to the upper parts of the SL1 stem-loops and opens their lower parts. The opening of the SL1 stem-loop leads the PAL self-complementary sequences (orange line) to form a duplex of 10 base pairs.*

4.1.5.2. *Dimerization of HIV-1 gRNA*

Dimerization of gRNA could be an early event, as the in vitro transcription of a DNA corresponding to the 5'UTR-gag region of the proviral DNA produces dimeric RNA (Darlix et al. 1990). Transcription of proviral DNA produces three types of unspliced RNAs with one, two or three guanines at the 5' end (Masuda et al. 2015). An RNA synthesized in vitro and corresponding to the first 344 nucleotides of the HIV-1 genome with two or three guanines at the 5' end exhibits a weak ability to dimerize, while one with one guanine dimerizes efficiently (Kharytonchyk et al. 2016). Two or three guanines at the 5' end of the viral genome alter its folding. Interestingly, it is the one-guanine unspliced RNA that is preferentially encapsidated in the virion as a gRNA dimer (Masuda et al. 2015; Kharytonchyk et al. 2016). Although unstable dimers can be generated in vitro in the absence of protein, it has not been shown that gRNA does not require protein to form the immature dimer. An ex vivo study suggests that the Gag precursor facilitates the formation of the immature dimer (Chen et al. 2016a).

An ex vivo study performed with retroviral vectors strongly suggests that dimerization of HIV-1 gRNA occurs prior to its encapsidation into the immature viral particle (Moore et al. 2007). A study based on total internal reflection fluorescence microscopy (usually abbreviated TIRF) suggests that dimeric gRNAs exist at the plasma membrane (Chen et al. 2016a). The presence of gRNA dimers in the cytosol and at the plasma membrane of living cells was demonstrated by a more precise study using a form of super-resolution microscopy called 3D-SIM and ccFCS, which is a form of fluorescence correlation spectroscopy (Ferrer et al. 2016). The latter study did not reveal gRNA dimers in the nucleus and is therefore in favor of gRNA dimerization being initiated in the cytosol; this does not exclude the formation of dimers at the plasma membrane. Analysis of RNA encapsidated in viral particles produced by fused cells also suggests that gRNA does not dimerize in the nucleus (Moore et al. 2009). However, the hypothesis of gRNA dimerization in the nucleus has not been completely abandoned because current analytical methods do not allow it to be disproved or confirmed beyond doubt.

Electron microscopy analysis of the gRNA dimer, which is extracted from the virion, indicates that the DLS is located within the first 400 nucleotides of the 5' end of the genome and consists of two contact points (Höglund et al. 1997). The same type of DLS has been observed with atomic force microscopy with an RNA dimer formed in vitro, with each subunit corresponding to the first 744 nucleotides of the 5' end of the HIV-1 genome (Andersen et al. 2004; Pallesen 2011). Results from studies focused on dimerization of in vitro synthesized RNAs suggest that the TAR and SL1 elements are the two contact points of the DLS observed by both microscopy techniques (see Figure 4.16).

Dimerization of HIV-1 gRNA has been the subject of numerous in vitro studies using RNAs containing DLS (see references cited in the study by Dubois et al. (2018)). The RNA, which is synthesized in vitro and corresponds to the TAR element, dimerizes efficiently when incubated in the presence of HIV-1 NC (Andersen et al. 2004). The apical portion of the TAR stem-loop has a self-complementary sequence (see Figure 4.17) that is responsible for dimerization.

Figure 4.16. *Sites involved in HIV-1 gRNA dimerization. The DLS identified by electron and atomic force microscopy is thought to involve the TAR and SL1 elements that are upstream of the splice donor site (SD). The orange disk represents the 5' cap. Regions and genes are not shown to scale. For a color version of this figure, see www.iste.co.uk/fosse/structures.zip*

In the context of a long RNA (744 nucleotides), the TAR element has only a minor role in NCp7-dependent dimerization in vitro (Andersen et al. 2004). Furthermore, studies of mutants suggest that ex vivo gRNA dimerization is not dependent on the TAR autocomplementary sequence (Jalalirad et al. 2012). Thus, the TAR–TAR interaction has not been shown to be one of the contact points that binds the gRNA dimer subunits in the virion. The primary and perhaps exclusive role of TAR is its ability to activate transcription of proviral DNA (Das et al. 2007).

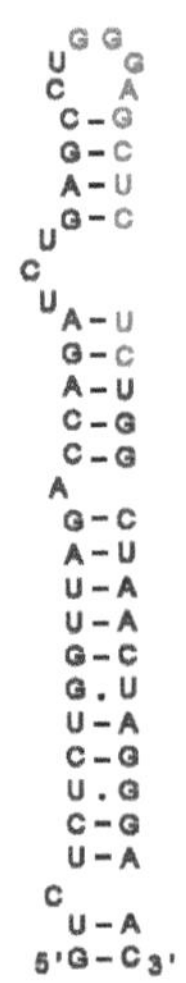

Figure 4.17. *Secondary structure of the TAR element. The secondary structure that is shown corresponds to that present in the gRNA monomer. The self-complementary sequence (10 nucleotides) initiating dimerization via the TAR element is in orange. For a color version of this figure, see www.iste.co.uk/fosse/structures.zip*

In the absence of the NC, the SL1 element has been identified as responsible for the dimerization of an in vitro synthesized RNA corresponding to the first 615 nucleotides of HIV-1 gRNA (Skripkin et al. 1994). The major role of the SL1 element in in vitro dimerization in the absence of an NC has been confirmed by other studies (Laughrea and Jetté 1994; Muriaux et al. 1995; Clever et al. 1996). The SL1 element forms a stem-loop structure, with the apical loop having a self-complementary sequence of six nucleotides (see Figure 4.18). In many HIV-1 isolates, there is strong selection pressure to maintain the autocomplementary sequences GUGCAC and GCGCGC in the apical loop (Hussein et al. 2010). Deletion of the stem-loop SL1 or mutations in its apical loop, which delete the autocomplementary sequence, decrease HIV-1 infectivity by a factor of 10 to 1,000 (Paillart et al. 1996b; Clever and Parslow 1997; Laughrea et al. 1997). Dimerization of gRNA in the cytosol and at the plasma membrane is significantly decreased by the deletion of the SL1 stem-loop or mutations in its apical loop (Ferrer et al. 2016). However, the gRNA extracted from virions is in the form of a dimer when the SL1 stem-loop is deleted or when its apical loop is mutated (Berkhout and van Wamel 1996; Clever and Parslow 1997). The gRNA can therefore form ex vivo a dimer that does not depend on the stem-loop SL1, but this dimer does not allow efficient replication of the virus. It has been proposed that the SL1 element initiates the formation of a gRNA dimer that has an optimal conformation for HIV-1 replication (Clever and Parslow 1997).

Under physiological temperature and salinity conditions, and in the absence of the NC, an RNA corresponding to the 5' part of the HIV-1 gRNA forms an unstable dimer in vitro via the SL1 stem-loop (Laughrea and Jetté 1996). This dimer may correspond in part to the gRNA dimer that is present in the immature viral particle. The subunits of this dimer dissociate when analyzed by agarose gel electrophoresis at room temperature in the absence of magnesium. The apical loop of the SL1 stem-loop is directly involved in the loop–loop interaction that binds the subunits of the unstable dimer (Clever et al. 1996; Paillart et al. 1996a). This interaction forms a short duplex of six base pairs (see Figure 4.18). Studies by NMR and X-ray crystallography of a short RNA (23–24 nucleotides) representing the upper part of the SL1 stem-loop have confirmed the loop–loop interaction (Ennifar et al. 2001; Kieken et al. 2006).

RNAs synthesized in vitro and representing the 5' end of the HIV-1 genome form stable dimers after incubation at elevated temperatures

(55–65°C) (Laughrea and Jetté 1996; Muriaux et al. 1996a). These in vitro studies strongly suggest that the subunits of the stable dimers are linked by a long duplex (28 base pairs) resulting from the opening and pairing of two SL1 stem-loops (see Figure 4.18). It was shown by NMR that a short RNA (35 nucleotides) corresponding to the entire SL1 sequence can form the long duplex (Ulyanov et al. 2006). A subsequent NMR study showed that an RNA synthesized in vitro and corresponding to the first 344 nucleotides of the 5' end of the gRNA can form a stable dimer at 37°C whose subunits are linked by the long duplex (Keane et al. 2016).

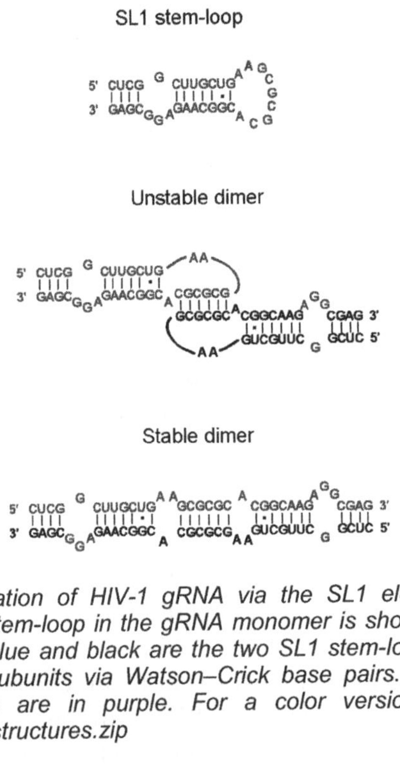

Figure 4.18. *Dimerization of HIV-1 gRNA via the SL1 element. The secondary structure of the SL1 stem-loop in the gRNA monomer is shown in the upper part of the figure. Shown in blue and black are the two SL1 stem-loops that link the stable and unstable dimer subunits via Watson–Crick base pairs. The internal loops on which the NC binds are in purple. For a color version of this figure, see www.iste.co.uk/fosse/structures.zip*

Under physiological conditions, in vitro formation of stable dimers is strongly stimulated by HIV-1 NC when RNAs contain the SL1 stem-loop (Muriaux et al. 1996b; Andersen et al. 2004). It is likely that the subunits of these dimers are associated by the long duplex (see Figure 4.18), but this has not been demonstrated. The conversion of the loop–loop complex to the long duplex has been studied in vitro with short RNAs (23 and 35 nucleotides) and different analytical techniques (gel electrophoresis, fluorescence, mass spectrometry and NMR) (Rist and Marino 2002; Hagan and Fabris 2007; Mujeeb et al. 2007). The results of these studies strongly suggest that the binding of NC on the inner loop destabilizes the apical stem of SL1 without dissociating the loop–loop complex, thus leading to the formation of the long duplex. To date, it has not been shown that the long duplex generated in vitro is the or one of the contact points that associates the subunits of the gRNA dimer in the mature virus particle.

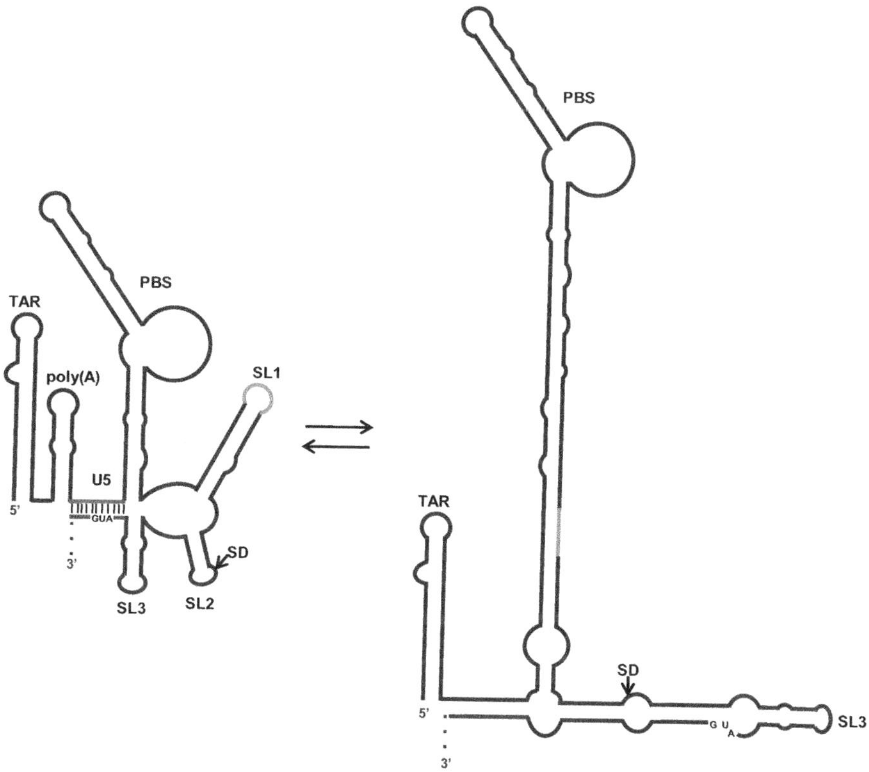

Figure 4.19. *First regulation model of HIV-1 gRNA dimerization by alternative folding of the 5'-UTR region. For a color version of this figure, see www.iste.co.uk/fosse/structures.zip*

COMMENTARY ON FIGURE 4.19.– *On the left is the BMH conformation, which can dimerize because the SL1 element forms a stem-loop. The self-complementary sequence (orange) is responsible for the loop–loop interaction in the unstable dimer. Eleven nucleotides of the sequence U5 (in red) are paired with the sequence (in green) that contains the initiation codon (AUG) of the gag gene. These two paired sequences constitute the beginning and the end of the encapsidation signal Ψ. On the right is shown the LDI conformation, which cannot dimerize because the SL1 element is paired in part with the poly(A) and U5 domains.*

RNAs corresponding to the 5'-UTR region and the beginning of the *gag* gene can adopt two alternative conformations in vitro (Abbink and Berkhout 2003). In the one called BMH, the SL1 element forms the stem-loop involved in dimerization, while in the other, called LDI, it interacts with the polyA and U5 elements and thus does not form the stem-loop (see Figure 4.19). It has been proposed that ex vivo the BMH conformation leads to gRNA dimerization and inhibits *gag* gene translation because the initiation codon of the *gag* gene is involved in a stable base pairing interaction that is conserved in different lentiviruses (Abbink and Berkhout 2003; Tran et al. 2015). Mutations that promote BMH folding, however, do not reduce translation ex vivo (Abbink et al. 2005). The secondary structure of the gRNA in the virion, which was determined using the SHAPE method (Watts et al. 2009), is consistent with the BMH conformation. It has been proposed that ex vivo the LDI conformation is competent for translation of the *gag* gene but not for gRNA dimerization. However, LDI folding has not been characterized ex vivo, and mutations that prevent its formation do not decrease translation ex vivo (Abbink et al. 2005).

A second alternative folding model of the 5'-UTR region and the beginning of the *gag* gene has been proposed from agarose gel electrophoresis and NMR analyses of an RNA corresponding to the 5' end of the HIV-1 genome (Lu et al. 2011). In this model, the BMH conformation, which is slightly different from the first model, is in equilibrium with the U5-SL1 conformation (see Figure 4.20). This BMH conformation is also consistent with the secondary structure of the gRNA in the virion, which was determined using the SHAPE method (Watts et al. 2009). It has been proposed that gRNA dimerization requires the BMH conformation, while translation of the *gag* gene is facilitated by the other conformation. It has not been shown ex vivo that U5-SL1 folding exists and promotes *gag* gene translation.

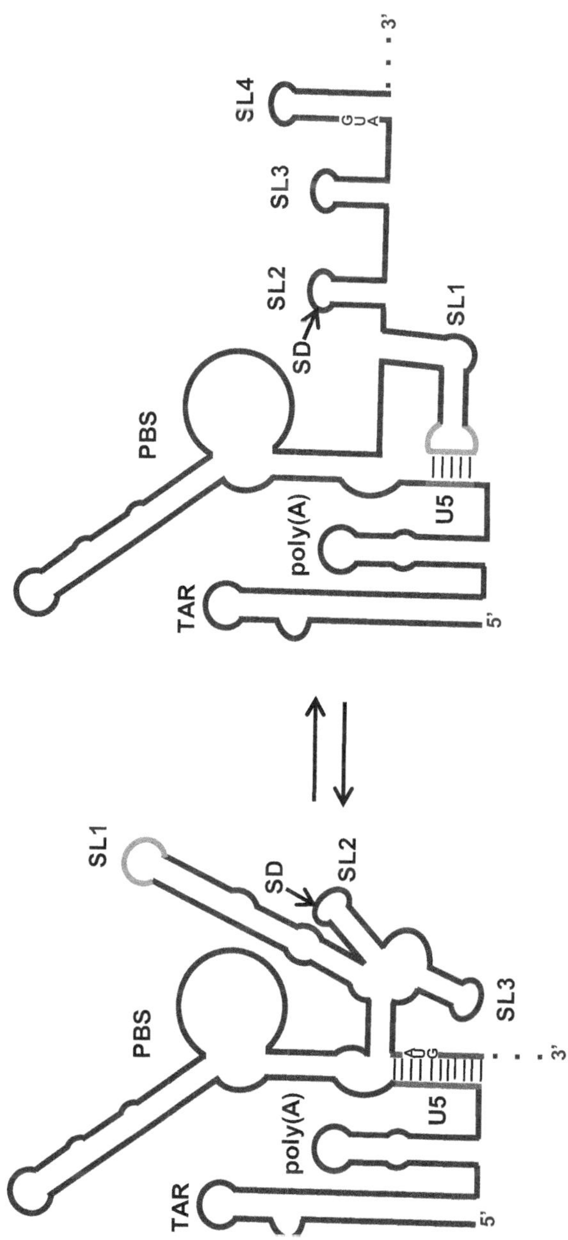

Figure 4.20. *Second regulation model of HIV-1 gRNA dimerization by alternative folding of the 5'UTR. For a color version of this figure, see www.iste.co.uk/fosse/structures.zip*

COMMENTARY ON FIGURE 4.20.– *On the left is shown the BMH conformation, which can dimerize because the SL1 element forms a stem-loop. The self-complementary sequence (orange) is responsible for the loop–loop interaction in the unstable dimer. Eleven nucleotides of the U5 sequence (in red) are paired with the sequence (in green) that contains the initiation codon (AUG) of the gag gene. These two paired sequences constitute the beginning and the end of the encapsidation signal Ψ. On the right is represented the U5-SL1 conformation, which cannot dimerize because five nucleotides of the autocomplementary sequence (six nucleotides) are paired with nucleotides of the U5 sequence. The initiation codon of the gag gene is in the SL4 stem-loop.*

4.2. RNA structures and mechanisms regulating gRNA encapsidation

In different retroviral species, the gRNA is recognized by the Gag precursor, usually through a specific interaction between the NC domain of the polypeptide and a structured RNA region that is called the encapsidation signal or domain. Less than 1% of Gag polypeptides (i.e. about a dozen molecules) are thought to be involved in gRNA recognition (Jouvenet et al. 2009; Miyazaki et al. 2010). The first studies focused on the identification of encapsidation signals of different retroviruses were performed with whole genomes (see references cited in the studies by Jouvenet et al. (2011) and Maldonado and Parent (2016)). They showed that the 5'-UTR region of retroviral gRNAs contains a major encapsidation signal. Following these initial studies, studies concerning the encapsidation of retroviral genomes were mainly performed with retroviral vectors for a simpler analysis of the results and a finer characterization of the RNA sequences and structures involved in gRNA encapsidation.

4.2.1. *Structures and mechanisms in alpharetroviruses*

Alpharetrovirus gRNA encapsidation has primarily been studied in RSV (Maldonado and Parent 2016). Although identical, RSV unspliced RNAs are thought to split into two populations in the nucleus cell (Maldonado and Parent 2016). The unspliced RNAs in one population serve as genomic

RNAs and interact with one or more Gag polypeptides to pass into the cytoplasm and then are directed to the plasma membrane where they are encapsidated into viral particles. It is not known whether it is a monomeric gRNA or an unstable gRNA dimer that interacts with one or more Gag precursors. The NC domain of the Gag polypeptide, mainly via its two zinc fingers, is responsible for the specific encapsidation of RSV gRNA (Méric and Spahr 1986; Méric et al. 1988). In addition, the formation of the gRNA dimer in the virion depends on the NC, which is released during the cleavage of the Gag precursor by the viral protease (Oertle and Spahr 1990; Stewart et al. 1990).

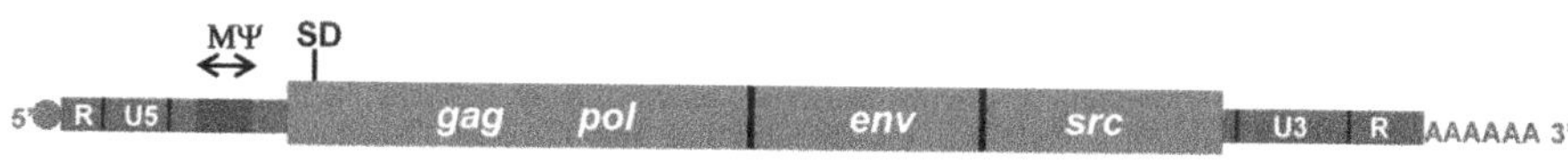

Figure 4.21. *RSV gRNA encapsidation signal. The MΨ encapsidation signal is upstream of SD and the gag gene. The orange disk represents the cap in 5'. Regions and genes are not shown to scale. For a color version of this figure, see www.iste.co.uk/fosse/structures.zip*

The MΨ sequence (160 nucleotides) is the main encapsidation signal of RSV because it contains much of the information necessary for gRNA encapsidation and leads to the encapsidation of non-viral RNA when fused to it. Thus, MΨ creates a 100-fold increase in non-viral RNA encapsidation (Aronoff et al. 1993; Banks et al. 1999). However, the encapsidation of a non-viral RNA containing MΨ is decreased by a factor of 3 compared to the whole gRNA (Banks et al. 1999). There may be a sequence or sequences downstream of MΨ that serve as secondary encapsidation signals, or that increase encapsidation by inducing a conformation of the gRNA that is optimal for the MΨ-Gag precursor interaction. These sequences present in the gRNA, downstream of the 250th nucleotide of the *gag* gene, have not been characterized. As the main RSV encapsidation signal is upstream of the SD (see Figure 4.21), it should also allow for selective and efficient encapsidation of viral spliced RNAs. However, analysis of RNAs encapsidated in virions shows that gRNA encapsidation is 15 times greater than that of *env* mRNA (Banks et al. 1999). One hypothesis is incorrect folding of the MΨ encapsidation signal into the *env* mRNA due to long-range interactions between sequences upstream of the SD and those present in the *env* gene.

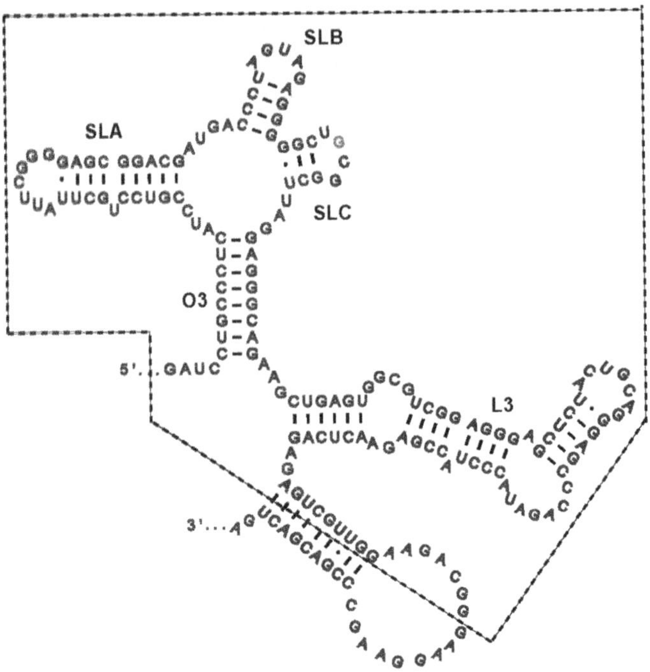

Figure 4.22. *Secondary structure of the MΨ encapsidation domain. For a color version of this figure, see www.iste.co.uk/fosse/structures.zip*

COMMENTARY ON FIGURE 4.22.– *The secondary structure of the MΨ domain (the dashed boxed portion) corresponds to that present in the RSV (Pr-C strain) gRNA monomer. Guanine 218 in orange is an important contact point in the Gag-MΨ interaction.*

The secondary structure of the gRNA in the virion and in the cell has not been determined. However, a recent study determined the secondary structure of an RNA synthesized in vitro corresponding to the first 636 nucleotides of the RSV genome (Liu et al. 2020). This in vitro study showed that the MΨ region is highly structured (see Figure 4.22) and that guanine 218 present in the apical loop of the SLC stem plays an important role in binding the Gag polypeptide to the MΨ domain. The results of this study combined with those of an NMR study (Zhou et al. 2007) strongly suggest that guanine 218 interacts with the N-terminal zinc finger of the NC domain of the Gag precursor. However, the NC domain of the Gag polypeptide has other contacts with MΨ to package the gRNA. Indeed, mutations in the apical loop of SLC only moderately decrease viral replication and the encapsidation of non-viral RNA containing MΨ (Banks and Linial 2000; Zhou et al. 2007). The MΨ region including the O3 stem and the SLA, SLB and SLC stem-loops plays a crucial role in encapsidation. Indeed, the deletion of the L3 stem-loop, which plays an essential role in RNA dimerization in vitro (see section 4.1.1), did not decrease the encapsidation of a non-viral RNA containing the other MΨ elements (Banks and Linial 2000).

4.2.2. *Structures and mechanisms in betaretroviruses*

Betaretroviruses, such as MPMV and MMTV, differ from other retroviral species in the assembly process of the immature viral particle, which takes place in the cytoplasm, not at the plasma membrane of the infected cell. It is during this process that the gRNA is encapsidated into the immature viral particle through its interaction with several Gag polypeptides. It is not known whether it is a monomeric gRNA or an unstable gRNA dimer that interacts with the polypeptides.

4.2.2.1. *Encapsidation of MPMV gRNA*

The specific gRNA encapsidation of MPMV depends on the C-terminus of the CA domain (RKK basic motif) and the NC domain of the Gag polypeptide (Füzik et al. 2016; Dostálková et al. 2018). Although not demonstrated for MPMV, it is likely that the activity of the NC domain results in part from its two zinc fingers. However, it has been shown that the KNKEK basic motif, which is just upstream of the N-terminal zinc finger (René et al. 2018), plays an important role in encapsidation (Dostálková et al. 2018).

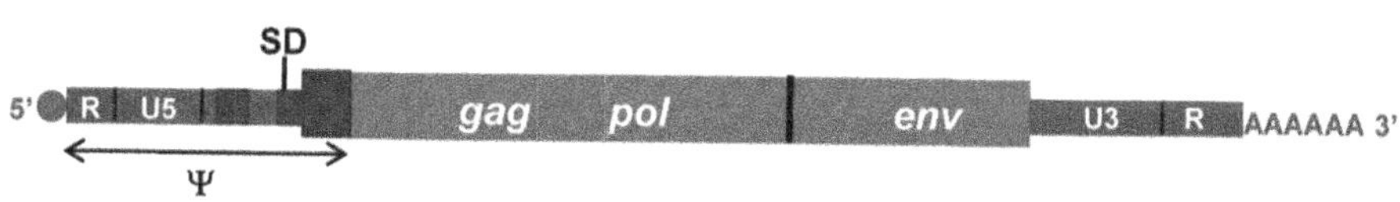

Figure 4.23. *MPMV gRNA encapsidation signal. The encapsidation signal is on either side of the splice donor site (SD). The two domains that are essential for the activity of Ψ are in red. The orange disk represents the 5' cap. Regions and genes are not shown to scale. For a color version of this figure, see www.iste.co.uk/fosse/structures.zip*

The use of retroviral vectors made it possible to identify the region named Ψ (389 nucleotides), which is the primary signal for MPMV gRNA encapsidation (Schmidt et al. 2003; Jaballah et al. 2010; Kalloush et al. 2016). The activity of Ψ is primarily due to two domains (see Figure 4.23). The first domain is just downstream of the PBS sequence and consists of 50 nucleotides. Its 3' end contains the Pal SL sequence that is involved in in vitro RNA dimerization (see section 4.2.1.1). This sequence is not sufficient for encapsidation, and it has not been shown to be required for ex vivo dimerization of gRNA. The second domain is composed of the last 23 nucleotides of the 5'-UTR region and the first 120 nucleotides of the *gag* gene. Encapsidation is reduced by 33 times when the first domain and the first 23 nucleotides of the second domain are deleted (Jaballah et al. 2010). Encapsidation is reduced threefold when the last 90 nucleotides of the second domain are deleted (Schmidt et al. 2003). The encapsidation of spliced RNAs is much less efficient than that of gRNA because they contain only the first four nucleotides of the second domain of Ψ.

The secondary structure of the gRNA in the virion and in the cell has not been characterized. The secondary structure of an RNA synthesized in vitro corresponding to the MPMV encapsidation signal has been determined (Aktar et al. 2013). This in vitro study showed that the region is highly structured (see Figure 4.24). The SL3, Gag SL1 and Gag SL2 stem-loops have minor roles in encapsidation (Kalloush et al. 2019). In contrast, the encapsidation of an RNA containing Ψ was greatly diminished when the two long-range interactions (LR-I and LR-II) between the U5 region and the *gag* gene were deleted by site-directed mutagenesis (Kalloush et al. 2016, 2019). LR-I and LR-II allow Ψ to adopt a conformation that is optimal for exposing the Gag polypeptide-interacting sites. In this conformation, two internal loops, which are accessible and located on either side of the Pal SL and SL3 stem-loops, constitute two high-affinity sites for the Gag precursor (Pitchai et al. 2021). These loops are rich in purines and GU motifs (sequence UUAAAAGUGAAAGUAA for one and sequence AAGUGU for the other). Since in all retroviruses except HTLV-1, NC binds preferentially to unpaired guanines and often to the GU motif (see references in René et al. (2018)), it is likely that on each loop a Gag precursor molecule binds via its NC domain. Interestingly, the *env* mRNA, which is only very weakly encapsidated, does not contain the Gag polypeptide binding site that is just downstream of the SD. It does have the site upstream of the SD, but this is partially paired and thus less accessible to the polypeptide (Pitchai et al. 2021).

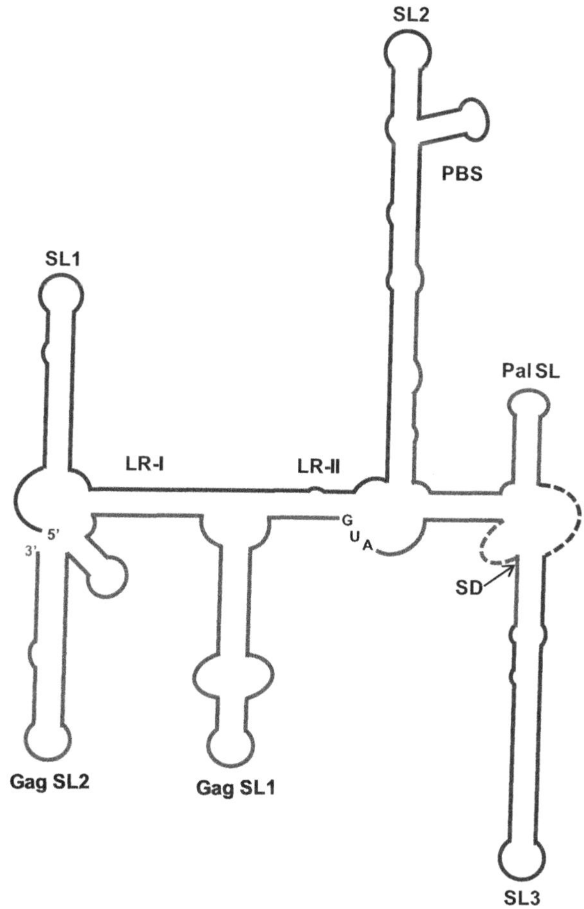

Figure 4.24. *Folding of the MPMV Ψ encapsidation signal. For a color version of this figure, see www.iste.co.uk/fosse/structures.zip*

COMMENTARY ON FIGURE 4.24.– *The folding of Ψ that is shown corresponds to that present in the MPMV gRNA monomer. The two domains that are primarily responsible for the activity of Ψ are in green. The two high-affinity sites for the Gag precursor are dashed. The splice donor site (SD) is at the 3′ end of the SL3 stem-loop. AUG is the initiation codon of the gag gene.*

4.2.2.2. *Encapsidation of MMTV gRNA*

Although not directly demonstrated by mutations targeting the Gag precursor, it is likely that MMTV gRNA encapsidation is dependent on this polypeptide. In favor of this hypothesis, mutations in the encapsidation signal strongly decrease encapsidation and Gag-binding in vitro (Chameettachal et al. 2021).

The use of retroviral vectors made it possible to identify the region named Ψ (432 nucleotides), which is the primary signal for MMTV gRNA encapsidation (Mustafa et al. 2012). This region contains the entire 5'-UTR domain and at least the first 120 nucleotides of the *gag* gene (see Figure 4.25).

Figure 4.25. *MMTV gRNA encapsidation signal. The encapsidation signal is on either side of the splice donor site (SD). The orange disk represents the 5' cap. Regions and genes are not shown to scale. For a color version of this figure, see www.iste.co.uk/fosse/structures.zip*

The secondary structure of the gRNA in the virion or in the cell has not been characterized. The secondary structure of an in vitro synthesized RNA corresponding to the MMTV encapsidation signal has been determined (Aktar et al. 2014). This in vitro study showed that the region is highly structured (see Figure 4.26). The long-range LRI interaction between the U5 region and the *gag* gene likely allows the Ψ to adopt a conformation that is optimal for exposing the Gag polypeptide-interacting sites. This hypothesis has not been supported by published results. Strong sites for Gag have not been identified in vitro in the 3' portion of the Ψ that corresponds to the first 120 nucleotides of the *gag* gene (Chameettachal et al. 2021). The structure of the SL4 element, consisting of a stem and two stem-loops, plays a key role in encapsidation. Indeed, encapsidation is decreased by a factor of 30 to 50 by mutations that suppress this structure (Mustafa et al. 2018). The apical loop of one of the two stem-loops is composed only of purines (GGAGAAGAG) and constitutes a strong binding site for the Gag precursor in vitro. The replacement of these purines by pyrimidines is deleterious to the encapsidation and binding of Gag to Ψ (Chameettachal et al. 2021). The sequence consisting of purines is also present in spliced viral RNAs, but being paired, it is not a high-affinity site for Gag.

The PBS element, which consists of a short stem-loop and an inner loop, is a second strong binding site for the Gag polypeptide. Mutations in the PBS element reduce encapsidation by a factor of 20 and significantly decrease the affinity of Gag for Ψ (Chameettachal et al. 2021). MMTV spliced RNAs contain the PBS element, which is no more paired than in gRNA and thus should possess a high-affinity site for Gag. However, it is not selectively encapsidated in the immature virus particle and weakly binds the Gag polypeptide (Chameettachal et al. 2021). This could be due to only one strong site being accessible to Gag (see above) and possibly also due to the low accessibility of the PBS element in spliced RNAs.

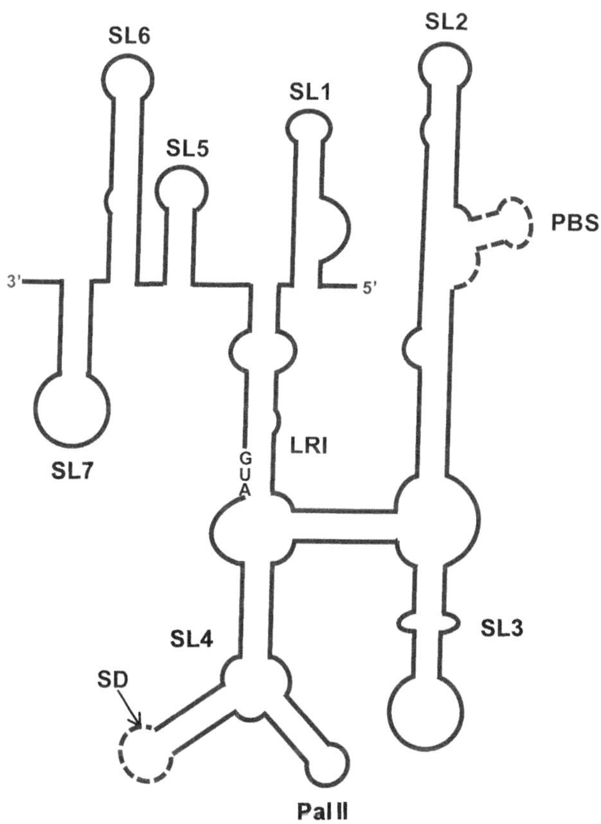

Figure 4.26. *Folding of the MMTV Ψ encapsidation signal*

COMMENTARY ON FIGURE 4.26.– *The folding of Ψ that is shown corresponds to that present in the MMTV gRNA monomer. The two high-affinity sites for the Gag precursor are dashed. The splice donor site (SD) is in the apical loop that binds Gag. Pal II is the primary site of RNA dimerization in vitro. AUG is the initiation codon of the* gag *gene.*

4.2.3. *Structures and mechanisms in deltaretroviruses*

The gRNA encapsidation of deltaretroviruses has been studied mainly in BLV and, as in all retroviral species, it requires the Gag polypeptide. However, it is interesting to note that in the case of BLV, encapsidation depends on the NC domain of Gag as well as on its MA domain. The two zinc fingers of the NC domain, as well as some basic amino acids of the MA domain, are probably involved in the Gag-signal encapsidation interaction (Wang et al. 2003). It is not known whether it is a monomeric gRNA or an unstable gRNA dimer that interacts with the Gag precursor.

Figure 4.27. *BLV gRNA encapsidation domains. For a color version of this figure, see www.iste.co.uk/fosse/structures.zip*

COMMENTARY ON FIGURE 4.27.– *The encapsidation signal consists of two domains (D1 and D2). D1 (147 nucleotides) is located in the 5'-UTR region downstream of the splice donor site (SD) and extends to the beginning of the* gag *gene. D2 (132 nucleotides) is in the central part of the* gag *gene. The orange disk represents the 5' cap. Regions and genes are not shown to scale.*

The use of retroviral vectors has shown that the encapsidation signal is discontinuous and composed of two domains that are downstream of the SD (Mansky et al. 1995). The first domain (D1), which plays the most important role in encapsidation, starts downstream of the PBS sequence and ends 70 nucleotides after the *gag* gene initiation codon (see Figure 4.27). The second domain (D2) is in the part of the *gag* gene that encodes the CA domain of the Gag polypeptide. The spliced viral RNAs are barely packaged because they do not contain D1 and D2. The secondary structure of the 5' end of the BLV gRNA has not been determined. Use of the mfold software showed

(Mansky and Wisniewski 1998) that both domains have the potential to form stem-loops (see Figure 4.28). Encapsidation is decreased fourfold by mutations that do not allow for the formation of a putative stem-loop (SL3) in D2 (Mansky and Wisniewski 1998). A portion of D1 encompassing the beginning of the *gag* gene has the potential to form two stem-loops (SL1 and SL2) that are largely responsible for encapsidation activity. Indeed, mutations that prevent the formation of these two stem-loops reduce encapsidation 40-fold (Mansky and Wisniewski 1998). Encapsidation is dependent on the sequence that is exposed in each apical loop (Mansky and Gajary 2002). Encapsidation of HTLV-1 gRNA, which is a deltaretrovirus, likely involves two stem-loops similar to SL1 and SL2. Indeed, a BLV-derived RNA can be encapsidated when both the SL1 and SL2 stem-loops of BLV are replaced by those of HTLV-1 (Mansky and Gajary 2002).

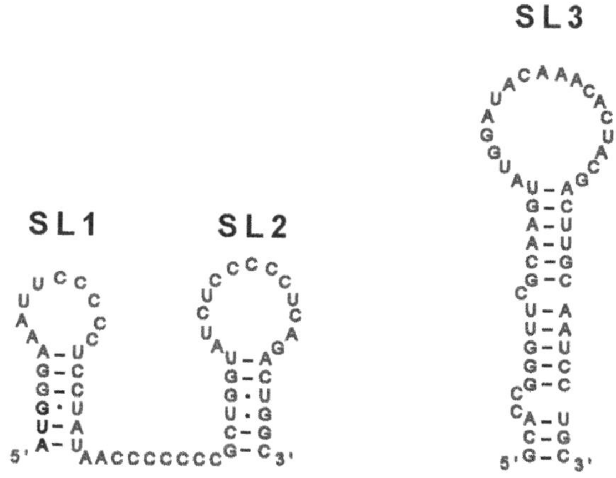

Figure 4.28. *Stem-loops involved in BLV gRNA encapsidation. Part of the D1 domain can form the SL1 and SL2 stem-loops, while part of the D2 domain can form the SL3 stem-loop. The initiation codon of the gag gene is in black. For a color version of this figure, see www.iste.co.uk/fosse/structures.zip*

The interaction between the BLV Gag polypeptide and its gRNA has not been studied. The stem-loop SL3 could interact with the NC domain of Gag because it contains three unpaired guanines. Indeed, the BLV NC, like the NCs of other retroviral species, binds preferentially to sequences containing unpaired guanines. In contrast, HTLV-1 NC does not have this specificity (Morcock et al. 2002). SL1 and SL2 are probably not high-affinity sites for the BLV Gag NC domain because they have only one unpaired guanine. The interaction of Gag with SL1 and SL2 depends primarily on its MA domain. In favor of this hypothesis, two isolated HTLV-1 stem-loops similar to SL1 and SL2 are two strong sites for HTLV-1 MA protein, while they do not bind HTLV-1 NC (Wu et al. 2018).

4.2.4. *Structures and mechanisms in gammaretroviruses*

The encapsidation of gammaretrovirus gRNA has been studied mainly in MoMuLV and, as in all retroviral species, it requires the Gag polypeptide. The zinc finger of the NC domain plays a crucial role in the Gag-signal encapsidation interaction (Gorelick et al. 1988; Berkowitz et al. 1995; Zhang and Barklis 1995).

Figure 4.29. *MoMuLV gRNA encapsidation domain. The encapsidation signal is located in the 5'-UTR region. It starts at the splice donor site (SD) and ends about 50 nucleotides from the gag gene initiation codon. The orange disk represents the 5' cap. Regions and genes are not shown to scale. For a color version of this figure, see www.iste.co.uk/fosse/structures.zip*

Several studies strongly suggest that the gRNA that is recognized by Gag is dimeric (Levin et al. 1974; Méric and Goff 1989; Hibbert et al. 2004). In vitro results suggest that some of the unspliced RNA molecules may dimerize during transcription of proviral DNA (Flynn and Telesnitsky 2006; Maurel et al. 2007). In addition, two ex vivo studies support the formation of unspliced RNA dimers in the nucleus (Flynn and Telesnitsky 2006; Maurel and Mougel 2010). The various microscopy techniques and strategies used to date have not been able to confirm or refute the presence of unspliced RNA dimers in the nucleus. The Gag polypeptide could be detected in the nucleus of MoMuLV-infected cells (Nash et al. 1993).

The Ψ region (350 nucleotides), which is located just downstream of the SD and upstream of the *gag* gene (see Figure 4.29), is the primary signal for MoMuLV gRNA encapsidation because its deletion suppresses this process (Mann et al. 1983). Although the region is sufficient for efficient encapsidation, the gRNA contains upstream and downstream sequences that promote optimal encapsidation (see references in the study by D'Souza and Summers (2005)).

A study conducted with chemical probes and RNA synthesized in vitro showed that the Ψ region forms an independent domain because its secondary structure is the same when isolated or present in the RNA corresponding to the first 725 nucleotides of the MoMuLV genome (Tounekti et al. 1992). This study suggests that Ψ forms 10 stem-loops in the gRNA monomer. Interestingly, these include the SLA, SLB, SLC and SLD stem-loops (see Figure 4.10), which are involved in gRNA dimerization (see section 4.1.4). Two ex vivo studies performed with retroviral vectors have shown that the SLC and SLD stem-loops are required for the encapsidation activity of Ψ (Mougel et al. 1996; Mougel and Barklis 1997).

The UCUG motif, which is present several times in the encapsidation signal, was identified in vitro as a site with high affinity for the NC (D'Souza and Summers 2004). Analysis by Weeks' team of the gRNA dimer, which is present in the virion, identified two high-affinity regions for NC in the encapsidation signal (Gherghe et al. 2010b). Each region is composed of two UCUG motifs (UCUG-UR-UCUG) and is unpaired in the gRNA dimers (see Figures 4.11 and 4.12). Interestingly, the UCUG-UR-UCUG sequence is present only in the gRNA encapsidation signal. Like NC, the Gag precursor binds preferentially to the unpaired UCUG motif (Gherghe et al. 2010b). The first U and G are the most important in the UCUG-Gag/NC interaction (Gherghe et al. 2010b). When the guanines of the four UCUG motifs in the encapsidation signal are replaced by adenines, the encapsidation of an RNA containing Ψ is decreased by a factor of 100 (Gherghe et al. 2010b). Studies by Weeks' team converge to propose a model in which each of the eight UCUG motifs in the immature gRNA dimer interacts with the NC domain of a Gag polypeptide (Grohman et al. 2014). The conformation adopted by Ψ in the dimeric gRNA is likely the product of an evolutionary process leading to optimal interactions between unpaired UCUG motifs and Gag polypeptides.

In vitro studies (Tounekti et al. 1992; D'Souza and Summers 2004; Gherghe et al. 2010a) suggest that the four UCUG motifs are not accessible in the gRNA monomer because they are at least partially paired (see Figure 4.10). Thus, dimerization of the gRNA is likely required for it to interact with multiple molecules of the Gag precursor. An appealing, but yet to be proven, hypothesis is that the Gag polypeptide recognizes in the nucleus the fraction of unspliced dimeric RNAs that thus serve as gRNAs, while the other fraction of unspliced monomeric RNAs are not recognized by the polypeptide but serve as mRNAs for the synthesis of Gag and Gag–Pol precursors.

4.2.5. *Structures and mechanisms in lentiviruses*

In lentiviruses, such as HIV-1 and HIV-2, the gRNA in the form of a dimer is encapsidated in the immature virus particle through its interaction with several Gag polypeptides (Moore et al. 2007; Ni et al. 2011; Chen et al. 2016a; Ferrer et al. 2016).

4.2.5.1. *Encapsidation of HIV-2 gRNA*

Although this has not been directly demonstrated by mutations targeting the Gag precursor, it is likely that HIV-2 gRNA encapsidation of HIV-2 gRNA is dependent on this polypeptide. In support of this hypothesis, the encapsidation of an RNA that corresponds in part to HIV-2 gRNA was significantly decreased when unpaired guanines, putatively interacting with the Gag precursor, were replaced by other nucleotides (Umunnakwe et al. 2021).

Three teams have studied HIV-2 gRNA encapsidation by analyzing the effect of deletions on both sides of the SD. These studies did not lead to a consensus on the location of the encapsidation signal. Indeed, it has been localized to the SD either upstream (McCann and Lever 1997; Griffin et al. 2001), downstream (Poeschla et al. 1998) or on both sides (Arya et al. 1998). The secondary structure of the gRNA in the virion and in the cell has not been characterized. The secondary structure of an RNA synthesized in vitro corresponding to the first 560 nucleotides of HIV-2 has been determined (Purzycka et al. 2011). This in vitro study showed that the 5'-UTR region and the beginning of the *gag* gene are highly structured (see Figure 4.30). Based on the assumption that in the overwhelming majority of retroviral species, the standard binding site for a Gag molecule is unpaired and contains at least one guanine, nine putative sites were tested by site-directed mutagenesis (Umunnakwe et al. 2021). These sites, corresponding to loops, contain 18 unpaired guanines (see Figure 4.30). Encapsidation is decreased by a factor of six when all of these guanines are replaced by other nucleotides. Sites six and nine apparently have no role in encapsidation.

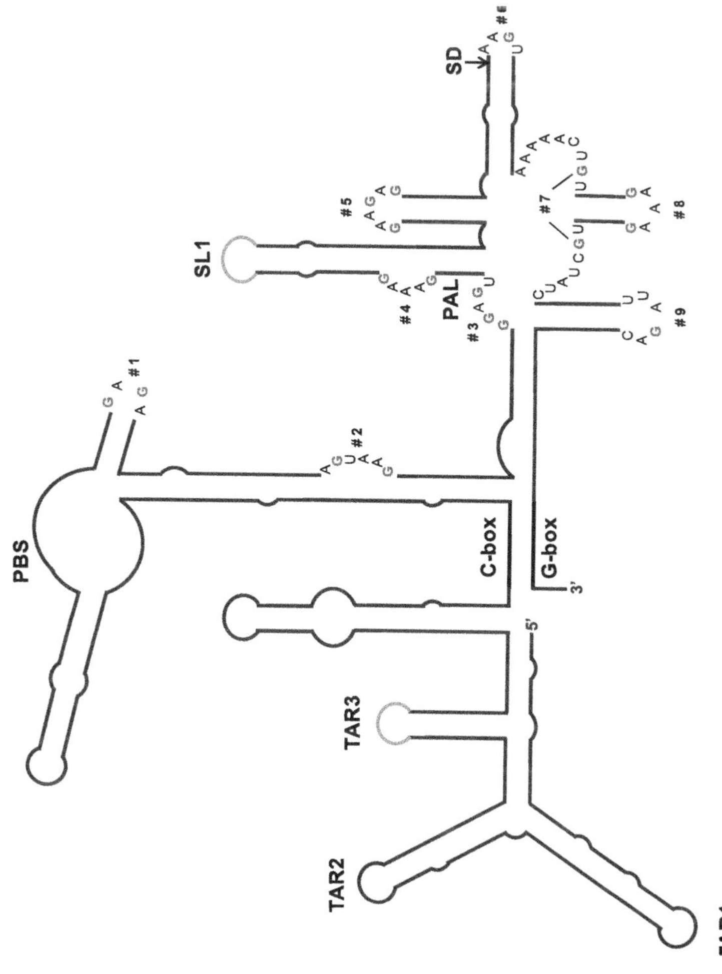

Figure 4.30. *Folding of the 5' end of the unstable HIV-2 gRNA dimer. For a color version of this figure, see www.iste.co.uk/fosse/structures.zip*

COMMENTARY ON FIGURE 4.30.– *The folding that is shown corresponds to that of a subunit of the unstable dimer formed by the RNA synthesized in vitro and corresponding to the 5' end of the gRNA. The two apical loops linking the subunits of the unstable dimer are in orange. Loop sequences are indicated where they have been mutated to determine their role in encapsidation. Guanines that have been replaced by other nucleotides are in red. With the exception of site #7, which consists of two internal loops, one site corresponds to a single loop. The C-box and G-box sequences (black lines) form an intramolecular base pairing interaction in the unstable dimer. The G-box sequence contains the initiation codon of the* gag *gene.*

In contrast, the other sites are involved in encapsidation and probably allow the binding of multiple Gag to the immature dimeric gRNA. The binding of multiple polypeptides to dimeric gRNA produces a synergistic effect that allows efficient encapsidation. Of the sites that can bind the Gag polypeptide, site 3 is the most important because replacement of its three guanines with other nucleotides decreases encapsidation by a factor of about two (L'Hernault et al. 2007; Baig et al. 2009; Umunnakwe et al. 2021).

4.2.5.2. *Encapsidation of HIV-1 gRNA*

The specific interaction of the Gag precursor via its two zinc fingers with the gRNA is necessary for the encapsidation of the latter into the virion (Aldovini and Young 1990; Gorelick et al. 1990). However, the two zinc fingers are not equivalent, with the N-terminal playing the most important role in encapsidation (Gorelick et al. 1993). The immature dimer gRNA that is present in the cytosol of HIV-1-infected cells (Ferrer et al. 2016) interacts with a limited number of Gag polypeptide molecules to be transported to the plasma membrane, where it is incorporated into the assembling virus particle (Kutluay and Bieniasz 2010; Hendrix et al. 2015). Association of the Gag polypeptide with unspliced viral RNA could be detected in the nucleus of HIV-1-infected cells (Ukah et al. 2018; Tuffy et al. 2020), but this association has not been shown to be the first step in the process of gRNA encapsidation (Grewe et al. 2012; Tuffy et al. 2020).

Deletions in the region between the SD and the initiation codon of the *gag* gene decrease gRNA encapsidation in the virion by at least a factor of 100 (Lever et al. 1989; Aldovini and Young 1990; Clavel and Orenstein 1990). Two studies (Heng et al. 2012; Kharytonchyk et al. 2018) suggest that the encapsidation signal is the region that lies between the R sequence and the first 15 nucleotides of the *gag* gene (see Figure 4.31). In these studies, the Ψ sequence (246 nucleotides) is sufficient to encapsidate a non-viral RNA when fused to it, which contains a signal that allows it to be transported from the nucleus to the cytoplasm. However, another study (Liu et al. 2017), performed with non-retroviral RNAs also containing a nuclear export signal, suggests that the domain is not sufficient to allow efficient encapsidation. This study also suggests that a significant portion (1,061 nucleotides) of the HIV-1 genome consisting of the entire 5'-UTR region and approximately the 5' half of the *gag* gene is required to allow optimal encapsidation of a non-retroviral RNA. Since the identification of the encapsidation signal was based on the analysis of deleted RNAs, it is likely that the above studies reached different conclusions because they did not use the same deletions or the same retroviral vectors. In the study by Hu's team (Liu et al. 2017), deletions upstream and downstream of Ψ could change its folding and thus indirectly inhibit its activity (Das et al. 2012; Kharytonchyk et al. 2018).

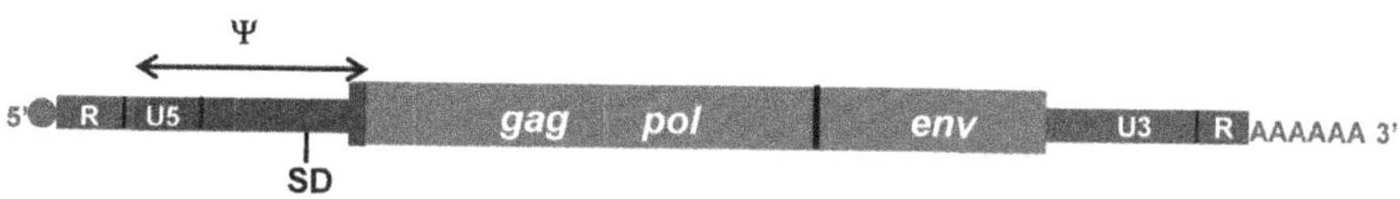

Figure 4.31. *HIV-1 gRNA encapsidation domain. The encapsidation signal is located almost exclusively in the 5'-UTR region. It begins downstream of the R sequence and ends 12 nucleotides after the gag gene initiation codon. The orange disk represents the 5' cap. Regions and genes are not shown to scale. For a color version of this figure, see www.iste.co.uk/fosse/structures.zip*

The secondary structure of the gRNA in the virion, which was determined by the SHAPE method (Watts et al. 2009), is consistent with the two BMH conformations that have been proposed for the 5'-UTR region (see Figures 4.19 and 4.20). The Ψ region folds into the form of the U5-AUG stem and the PBS, SL1, SL2 and SL3 stem-loops. The U5-AUG, SL1 and SL3 structural elements play an important role in the activity of Ψ, whereas the PBS and SL2 stem-loops are not directly involved (McBride and Panganiban 1997; Houzet et al. 2007; Heng et al. 2012).

Using the SHAPE method and an inhibitor that inhibits the binding of NC to RNA, it was possible to identify seven putative NC sites in the Ψ region of the dimeric gRNA that is in the virion (Wilkinson et al. 2008). These sites likely correspond to sites recognized by Gag via its NC domain. The sites all have one to three unpaired guanines and are downstream of the PBS stem-loop (see Figure 4.32). The unpaired guanines in the putative Gag sites were replaced with other nucleotides to study the effect of these substitutions on encapsidation (Nikolaitchik et al. 2020). The guanines in the apical loop of SL2 were not mutated because they overlap with the SD. The unpaired guanines act synergistically in encapsidation. Indeed, encapsidation is minimally decreased by substitutions located in two putative Gag sites, whereas it is reduced by a factor of 7 to 13 when four putative Gag sites are mutated, and by a factor of 34 when all putative Gag sites are mutated except the one containing the SD. Of the six sites studied, the two that are located in the lower part of the SL1 stem-loop are the most important for encapsidation. Interestingly, these two sites interact strongly with the Gag precursor in vitro (Abd El-Wahab et al. 2014). Taken together, these results suggest that binding a Gag polypeptide to each of the seven sites produces a synergistic effect that allows for optimal encapsidation of immature dimeric gRNA.

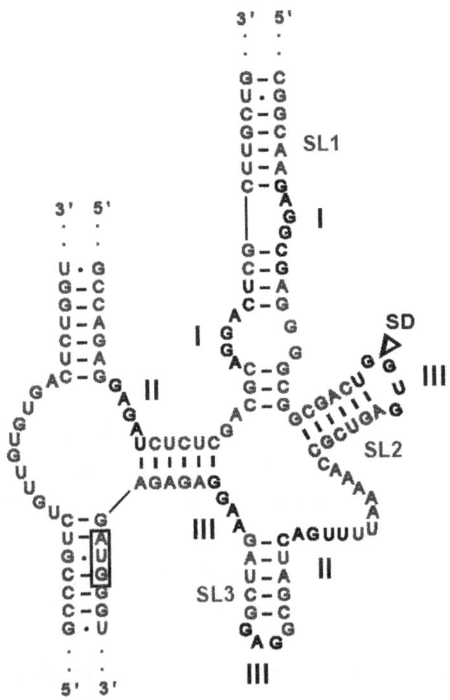

Figure 4.32. *Secondary structure of a Ψ portion of HIV-1. For a color version of this figure, see www.iste.co.uk/fosse/structures.zip*

COMMENTARY ON FIGURE 4.32.– *The secondary structure that is shown corresponds to a portion of Ψ in one subunit of the stable dimer that is present in the virion. The seven putative Gag sites are in black, and they are annotated I, II and III according to their strength (strong, medium and weak). The PBS stem-loop upstream of annotated site II is not shown, nor is the SL1–SL1 interaction linking the two dimer subunits. Only the lower part of the SL1 stem-loop, containing two putative strong binding sites for Gag, is visible in this secondary structure model. The initiation codon of the* gag *gene is boxed.*

Conclusion

As shown in this book through its four chapters and the literature cited, the study of structure–function relationships of retroviral RNAs has been energized over the last three decades by research on HIV-1, which is the main causative agent of the AIDS pandemic. During this period, new molecular biology technologies based on retroviral vectors, imaging and the SHAPE method, among others, have made it possible to determine ex vivo and in vitro the secondary structures of RNAs associated with specific functions. These structures are distributed in the coding and non-coding regions of the retroviral genome. The 5'-UTR non-coding regions of retroviruses contain the most secondary structures with functional roles.

The retroviral 5'-UTR is highly structured, and some of its structural elements perform different functions at different stages of the retroviral cycle. The Psi domain of MoMuLV, which is in the 5'-UTR and consists of four stem-loops, is a prominent example. Indeed, this domain has three functions: nuclear export of unspliced RNA, dimerization and encapsidation of gRNA. The secondary structures of the entire 5'-UTR regions of several retroviruses have been well-characterized, but the three-dimensional structures have not. Structural studies based on improvements of existing tools (NMR and cryo-electron microscopy) and technological innovations are needed to determine the tertiary structure of a 5'-UTR region at good resolution. To elucidate the molecular basis of a function associated with a 5'-UTR region during a specific step of the retroviral cycle, structural analysis of the 5'-UTR region will have to be performed in the presence of the viral/cellular proteins required for that step. In addition, the folding dynamics of RNA regions and their potential alternative folds must be taken into account to accurately determine the structure–function relationships.

Structure-dependent functions are based on RNA–RNA (tRNA–PBS, gRNA dimerization), DNA–RNA (first strand transfer and strand transfers responsible for recombination events) and RNA–protein (TAR–Tat, RRE–Rev and Psi–Gag in the case of HIV-1) interactions. The secondary structures responsible for these functions are generally involved in intermolecular interactions via unpaired and accessible nucleotides, such as the loop–loop interaction in gRNA dimerization, the TAR–Tat interaction in HIV-1 proviral DNA transcription and the Psi–Gag interaction in gRNA encapsidation.

Although a wealth of knowledge about the structure–function relationships of retroviral RNAs has been accumulated over the past three decades, much remains to be discovered about the RNA-dependent molecular mechanisms that regulate several steps of the retroviral cycle. To address the emergence of antiretroviral resistance, it is important to have several families of compounds that inhibit HIV-1 replication. To date, there is no clinically used antiretroviral that inhibits HIV-1 replication by specifically interacting with an RNA structure with a function. Further research and knowledge are therefore required to design such an antiretroviral.

Glossary

Alternative splicing: the splicing that produces different mRNAs from the same pre-messenger RNA.

Amino acid: an organic compound possessing a carboxyl group (–COOH) and a primary ($–NH_2$) or secondary (–NH–) amine group. Amino acids are the basic constituents of proteins.

Apical loop: a loop at the end of a stem-loop.

Branch point: a conserved sequence in the intron that contains an adenosine and plays an essential role in splicing.

Bulge: a loop on only one of the two strands of a stem.

Cap: methylated guanosine in position N7 that is linked to the first nucleotide of the mRNA by a 5'–5' triphosphate linkage. This is necessary for the protection of mRNA and its translation by the ribosome.

Chemical probe: this is a chemical agent that enables the identification of paired and unpaired nucleotides in the nucleic acid studied. The identification of the nucleotides modified by the chemical agent is performed by primer extension by means of a reverse transcriptase (indirect method) or by the use of RNA marked at one end (direct method).

Cis-regulatory element: a stimulatory or inhibitory sequence that acts on the DNA or RNA molecule that contains it.

Clinical isolate: an isolate from a patient.

Coaxial stacking: a process in which two RNA helices form a contiguous helix that is stabilized by the stacking of bases at the interface of the two helices.

Complex retroviruses: retroviruses whose genome also codes for accessory and regulatory proteins.

Crm1 pathway: the nuclear export pathway requiring the Crm1 protein, also called exportin 1 (XPO1). The activity of Crm1 depends on its association with Ran-GTP, which is a small GTP-bound G protein. Crm1–Ran^{GTP} transports a very large number of proteins from the nucleus to the cytoplasm by recognizing the NES they possess.

Encapsidation: the process by which the retrovirus genomic RNA is incorporated into the virus particle.

Enzymatic probe: an enzyme (nuclease) that allows us to identify the paired and unpaired nucleotides in the accessible regions of a nucleic acid. The identification of the enzyme cleavage sites is done by primer extension by means of a reverse transcriptase (indirect method) or through the use of RNA marked at one end (direct method).

Ex vivo: experiments performed on cells in culture.

Exon: the part of the gene that is retained after splicing. It is usually coding.

hnRNP proteins: proteins classified into different groups (A, H, etc.) that are present in heterogeneous ribonucleoprotein particles consisting of proteins and pre-messenger RNA. They are involved in alternative splicing.

In vitro: experiments performed in test tubes to reproduce biological processes.

In vivo: experiments performed on a living organism.

Internal loop: a loop that connects two stems.

Intron: the part of the gene that is not retained after splicing. It is generally non-coding.

Isolate: a virus isolated from a host and well-characterized genetically. It reproduces in cell culture.

Kilobase: the unit of measurement corresponding to 1,000 nucleotide bases.

Loop: a sequence of unpaired nucleotides in an RNA secondary structure.

Multiple loop: a loop that connects three or more stems.

NXF1/NXT1 pathway: the main pathway of nuclear export of cellular mRNAs involving the heterodimer formed by the proteins NXF1 (also called Tap) and NXT1 (also called p15).

Oncogene: a gene whose expression promotes carcinogenesis.

Primer: a short strand of DNA or RNA complementary to a part of a template that a DNA polymerase uses to start DNA synthesis by copying this template.

Secondary structure: the secondary structure of a single-stranded DNA/RNA molecule describes base pairings (Watson–Crick and G-U type) and unpaired bases. A secondary structure consists of stems, stem-loops, bulges, apical loops, inner loops and multiple loops. Numerous secondary structures of various RNAs have been constructed from the experimental data obtained with the chemical and enzymatic probes.

SHAPE: a method that makes it possible to determine the secondary structure of RNA. It is based on a chemical agent that acylates the 2'-OH group of a nucleotide when the latter is flexible; in general, this corresponds to an unpaired nucleotide. The identification of nucleotides modified by the acylating agent is performed by primer extension using a reverse transcriptase.

Simple retroviruses: retroviruses whose genome codes only for structural proteins and the three viral enzymes (IN, PR and RT).

Spliceosome: a large complex of proteins and small nuclear RNAs that is responsible for splicing pre-messenger RNAs.

Splicing: the maturation process of the pre-messenger RNA that eliminates introns and joins exons. It is based on the reactions involving cutting and ligation of the pre-messenger RNA strand.

SR proteins: proteins rich in arginines and lysines that bind to pre-messenger RNAs and are required for various steps of the spliceosome assembly.

Stem: a paired region that forms a helix.

Stem-loop: a hairpin structure consisting of a stem with one end closed by a loop called an apical loop.

Template: a strand of nucleic acid that is copied by a polymerase.

Tertiary interaction: a long-distance interaction that participates in the formation of the three-dimensional structure of RNA.

UBA: a protein domain consisting of about 40 amino acids that binds ubiquitin. This domain has been proposed to limit ubiquitin chain elongation and to direct ubiquitinated proteins to degradation via the 26S proteasome.

Virion: a mature and infectious viral particle.

List of Acronyms

3D-SIM	three-dimensional super-resolution structured illumination microscopy
AIDS	acquired immunodeficiency syndrome
ALV	avian leukosis virus
BIV	bovine immunodeficiency virus
CA	capsid protein
CAE	cytoplasmic accumulation element
ccFCS	cross-correlation fluctuation correlation spectroscopy
cDNA	complementary DNA
Crm1	chromosome region maintenance 1
CRS	cis-active repressive sequences
CTE	constitutive transport element
DIS	dimer initiation site
DLS	dimer linkage structure
DNA	deoxyribonucleic acid

DR	direct repeat
EIAV	equine infectious anemia virus
eIF4E	eukaryotic translation initiation factor 4e
F-MLV	Friend murine leukemia virus
gRNA	genomic RNA (of the retrovirus)
HIV-1	human immunodeficiency virus type 1
HIV-2	human immunodeficiency virus type 2
hnRNP A1	heterogeneous nuclear ribonucleoprotein A1
hRIP	human Rev-interacting protein
HTLV-1	human T-cell leukemia virus type 1
HTLV-2	human T-cell leukemia virus type 2
Hu	Hu antigen R
IN	integrase
IRES	internal ribosome entry site
ITAF	IRES trans-acting factors
JDV	Jembrana disease virus
JSRV	Jaagsiekte sheep retrovirus
kb	kilobase
L	leader (the sequence between the PBS and the *gag* gene)
LRR	leucine-rich repeat
LTR	long terminal repeat

MA	matrix protein
MLV	murine leukemia virus
MMTV	mouse mammary tumor virus
MoMuLV	Moloney murine leukemia virus
MPMV	Mason–Pfizer monkey virus
mRNA	messenger RNA
NC	nucleocapsid protein
NES	nuclear export signal
NLS	nuclear localization signal
NMR	nuclear magnetic resonance
NPC	nuclear pore complex
NRS	negative regulator of splicing
NTF2	nuclear transport factor 2
NTF2L	NTF2-like domain
NXF1	nuclear RNA export factor 1
NXT1	NTF2-related export protein 1
PBS	primer binding site
PPT	polypurine tract
PR	viral protease
Psi	packaging signal
PTB	polypyrimidine tract-binding protein

PTE	post-transcriptional element
R	repeat (the repeated sequence in the 5' and 3' of the retroviral genome)
Rev	regulator of expression of virion proteins
REV-A	reticuloendotheliosis virus type A
RmRE	RNA element required for Rem responsiveness
RNA	ribonucleic acid
RRE	Rev responsive element
RRM	RNA recognition motif
RSL	R region stem-loop
RSV	Rous sarcoma virus
RT	reverse transcriptase
RxRE	Rex responsive element
SAXS	small-angle X-ray scattering
SHAPE	selective 2'-hydroxylacylation analyzed by primer extension
SIV	simian immunodeficiency virus
SL	stem-loop
SRV-1	simian retrovirus type 1
ssDNA	strong-stop DNA
SU	surface protein
Tap	Tip-associated protein

TAR	transactivator response element
Tat	transactivator of transcription
TIM-TAM	Tat IRES modulator of Tat mRNA
TM	transmembrane protein
TRBP	TAR RNA binding protein
TREX1	transcription–export complex 1
tRNA	transfer RNA
U2AF	U2 auxiliary factor
U3	unique 3' sequence of the retroviral genome
U5	unique 5' sequence of the retroviral genome
UBA	ubiquitin-associated domain
UPF1	up-frameshift 1
UTR	untranslated region

References

Aagaard, L., Rasmussen, S.V., Mikkelsen, J.G., Pedersen, F.S. (2004). Efficient replication of full-length murine leukemia viruses modified at the dimer initiation site regions. *Virology*, 318, 360–370.

Abbink, T.E.M. and Berkhout, B. (2003). A novel long distance base-pairing interaction in human immunodeficiency virus type 1 RNA occludes the Gag start codon. *J. Biol. Chem.*, 278, 11601–11611.

Abbink, T.E.M., Ooms, M., Haasnoot, P.C.J., Berkhout, B. (2005). The HIV-1 leader RNA conformational switch regulates RNA dimerization but does not regulate mRNA translation. *Biochemistry*, 44, 9058–9066.

Abd El-Wahab, E.W., Smyth, R.P., Mailler, E., Bernacchi, S., Vivet-Boudou, V., Hijnen, M., Jossinet, F., Mak, J., Paillart, J.-C., Marquet, R. (2014). Specific recognition of the HIV-1 genomic RNA by the Gag precursor. *Nat. Commun.*, 5, 4304.

Aiyar, A., Cobrinik, D., Ge, Z., Kung, H.J., Leis, J. (1992). Interaction between retroviral U5 RNA and the T psi C loop of the tRNA(Trp) primer is required for efficient initiation of reverse transcription. *J. Virol.*, 66, 2464–2472.

Aktar, S.J., Jabeen, A., Ali, L.M., Vivet-Boudou, V., Marquet, R., Rizvi, T.A. (2013). SHAPE analysis of the 5' end of the Mason-Pfizer monkey virus (MPMV) genomic RNA reveals structural elements required for genome dimerization. *RNA*, 19, 1648–1658.

Aktar, S.J., Vivet-Boudou, V., Ali, L.M., Jabeen, A., Kalloush, R.M., Richer, D., Mustafa, F., Marquet, R., Rizvi, T.A. (2014). Structural basis of genomic RNA (gRNA) dimerization and packaging determinants of mouse mammary tumor virus (MMTV). *Retrovirology*, 11, 96.

Aldovini, A. and Young, R.A. (1990). Mutations of RNA and protein sequences involved in human immunodeficiency virus type 1 packaging result in production of noninfectious virus. *J. Virol.*, 64, 1920–1926.

Allain, B., Lapadat-Tapolsky, M., Berlioz, C., Darlix, J.L. (1994). Transactivation of the minus-strand DNA transfer by nucleocapsid protein during reverse transcription of the retroviral genome. *EMBO J.*, 13, 973–981.

Anand, K., Schulte, A., Vogel-Bachmayr, K., Scheffzek, K., Geyer, M. (2008). Structural insights into the cyclin T1-Tat-TAR RNA transcription activation complex from EIAV. *Nat. Struct. Mol. Biol.*, 15, 1287–1292.

Andersen, E.S., Contera, S.A., Knudsen, B., Damgaard, C.K., Besenbacher, F., Kjems, J. (2004). Role of the trans-activation response element in dimerization of HIV-1 RNA. *J. Biol. Chem.*, 279, 22243–22249.

Arhel, N.J., Souquere-Besse, S., Munier, S., Souque, P., Guadagnini, S., Rutherford, S., Prévost, M.-C., Allen, T.D., Charneau, P. (2007). HIV-1 DNA Flap formation promotes uncoating of the pre-integration complex at the nuclear pore. *EMBO J.*, 26, 3025–3037.

Aronoff, R., Hajjar, A.M., Linial, M.L. (1993). Avian retroviral RNA encapsidation: Reexamination of functional 5' RNA sequences and the role of nucleocapsid Cys-His motifs. *J. Virol.*, 67, 178–188.

Arrigo, S. and Beemon, K. (1988). Regulation of Rous sarcoma virus RNA splicing and stability. *Mol. Cell. Biol.*, 8, 4858–4867.

Arya, S.K., Zamani, M., Kundra, P. (1998). Human immunodeficiency virus type 2 lentivirus vectors for gene transfer: Expression and potential for helper virus-free packaging. *Hum. Gene. Ther.*, 9, 1371–1380.

Askjaer, P. and Kjems, J. (1998). Mapping of multiple RNA binding sites of human T-cell lymphotropic virus type I rex protein within 5'- and 3'-Rex response elements. *J. Biol. Chem.*, 273, 11463–11471.

Azoulay, J., Clamme, J.P., Darlix, J.L., Roques, B.P., Mély, Y. (2003). Destabilization of the HIV-1 complementary sequence of TAR by the nucleocapsid protein through activation of conformational fluctuations. *J. Mol. Biol.*, 326, 691–700.

Bai, Y., Tambe, A., Zhou, K., Doudna, J.A. (2014). RNA-guided assembly of Rev-RRE nuclear export complexes. *Elife*, 3, e03656.

Baig, T.T., Lanchy, J.-M., Lodmell, J.S. (2009). Randomization and in vivo selection reveal a GGRG motif essential for packaging human immunodeficiency virus type 2 RNA. *J. Virol.*, 83, 802–810.

Balakrishnan, M., Roques, B.P., Fay, P.J., Bambara, R.A. (2003). Template dimerization promotes an acceptor invasion-induced transfer mechanism during human immunodeficiency virus type 1 minus-strand synthesis. *J. Virol.*, 77, 4710–4721.

Baltimore, D. (1970). RNA-dependent DNA polymerase in virions of RNA tumour viruses. *Nature*, 226, 1209–1211.

Banks, J.D. and Linial, M.L. (2000). Secondary structure analysis of a minimal avian leukosis-sarcoma virus packaging signal. *J. Virol.*, 74, 456–464.

Banks, J.D., Kealoha, B.O., Linial, M.L. (1999). An MPSI-containing heterologous RNA, but not env mRNA, is efficiently packaged into avian retroviral particles. *J. Virol.*, 73, 8926–8933.

Bar, A., Marchand, V., Khoury, G., Dreumont, N., Mougin, A., Robas, N., Stévenin, J., Visvikis, A., Branlant, C. (2011). Structural and functional analysis of the Rous sarcoma virus negative regulator of splicing and demonstration of its activation by the 9G8 SR protein. *Nucleic Acids Res.*, 39, 3388–3403.

Barat, C., Lullien, V., Schatz, O., Keith, G., Nugeyre, M.T., Grüninger-Leitch, F., Barré-Sinoussi, F., LeGrice, S.F., Darlix, J.L. (1989). HIV-1 reverse transcriptase specifically interacts with the anticodon domain of its cognate primer tRNA. *EMBO J.*, 8, 3279–3285.

Barraud, P., Gaudin, C., Dardel, F., Tisné, C. (2007). New insights into the formation of HIV-1 reverse transcription initiation complex. *Biochimie*, 89, 1204–1210.

Barrera, A., Olguín, V., Vera-Otarola, J., López-Lastra, M. (2020a). Cap-independent translation initiation of the unspliced RNA of retroviruses. *Biochim. Biophys. Acta. Gene. Regul. Mech.*, 1863, 194583.

Barrera, A., Ramos, H., Vera-Otarola, J., Fernández-García, L., Angulo, J., Olguín, V., Pino, K., Mouland, A.J., López-Lastra, M. (2020b). Post-translational modifications of hnRNP A1 differentially modulate retroviral IRES-mediated translation initiation. *Nucleic Acids Res.*, 48, 10479–10499.

Bartels, H. and Luban, J. (2014). Gammaretroviral pol sequences act in cis to direct polysome loading and NXF1/NXT-dependent protein production by gag-encoded RNA. *Retrovirology*, 11, 73.

Baskerville, S., Zapp, M., Ellington, A.D. (1995). High-resolution mapping of the human T-cell leukemia virus type 1 Rex-binding element by in vitro selection. *J. Virol.*, 69, 7559–7569.

Basu, V.P., Song, M., Gao, L., Rigby, S.T., Hanson, M.N., Bambara, R.A. (2008). Strand transfer events during HIV-1 reverse transcription. *Virus. Res.*, 134, 19–38.

Basyuk, E., Boulon, S., Skou Pedersen, F., Bertrand, E., Vestergaard Rasmussen, S. (2005). The packaging signal of MLV is an integrated module that mediates intracellular transport of genomic RNAs. *J. Mol. Biol.*, 354, 330–339.

Baudin, F., Marquet, R., Isel, C., Darlix, J.L., Ehresmann, B., Ehresmann, C. (1993). Functional sites in the 5' region of human immunodeficiency virus type 1 RNA form defined structural domains. *J. Mol. Biol.*, 229, 382–397.

Belfetmi, A., Zargarian, L., Tisné, C., Sleiman, D., Morellet, N., Lescop, E., Maskri, O., René, B., Mély, Y., Fossé, P. et al. (2016). Insights into the mechanisms of RNA secondary structure destabilization by the HIV-1 nucleocapsid protein. *RNA*, 22, 506–517.

Ben Ali, M., Chaminade, F., Kanevsky, I., Ennifar, E., Josset, L., Ficheux, D., Darlix, J.-L., Fossé, P. (2007). Structural requirements for nucleocapsid protein-mediated dimerization of avian leukosis virus RNA. *J. Mol. Biol.*, 372, 1082–1096.

Bender, W. and Davidson, N. (1976). Mapping of poly (A) sequences in the electron microscope reveals unusual structure of type C oncornavirus RNA molecules. *Cell*, 7, 595–607.

Bender, W., Chien, Y.H., Chattopadhyay, S., Vogt, P.K., Gardner, M.B., Davidson, N. (1978). High-molecular-weight RNAs of AKR, NZB, and wild mouse viruses and avian reticuloendotheliosis virus all have similar dimer structures. *J. Virol.*, 25, 888–896.

Berkhout, B. and van Wamel, J.L. (1996). Role of the DIS hairpin in replication of human immunodeficiency virus type 1. *J. Virol.*, 70, 6723–6732.

Berkhout, B., Vastenhouw, N.L., Klasens, B.I., Huthoff, H. (2001). Structural features in the HIV-1 repeat region facilitate strand transfer during reverse transcription. *RNA*, 7, 1097–1114.

Berkowitz, R.D., Ohagen, A., Höglund, S., Goff, S.P. (1995). Retroviral nucleocapsid domains mediate the specific recognition of genomic viral RNAs by chimeric Gag polyproteins during RNA packaging in vivo. *J. Virol.*, 69, 6445–6456.

Berlioz, C. and Darlix, J.L. (1995). An internal ribosomal entry mechanism promotes translation of murine leukemia virus gag polyprotein precursors. *J. Virol.*, 69, 2214–2222.

Bernacchi, S., Stoylov, S., Piémont, E., Ficheux, D., Roques, B.P., Darlix, J.L., Mély, Y. (2002). HIV-1 nucleocapsid protein activates transient melting of least stable parts of the secondary structure of TAR and its complementary sequence. *J. Mol. Biol.*, 317, 385–399.

Bieth, E., Gabus, C., Darlix, J.L. (1990). A study of the dimer formation of Rous sarcoma virus RNA and of its effect on viral protein synthesis in vitro. *Nucleic Acids Res.*, 18, 119–127.

Blain, S.W. and Goff, S.P. (1995). Effects on DNA synthesis and translocation caused by mutations in the RNase H domain of Moloney murine leukemia virus reverse transcriptase. *J. Virol.*, 69, 4440–4452.

Bogerd, H.P., Huckaby, G.L., Ahmed, Y.F., Hanly, S.M., Greene, W.C. (1991). The type I human T-cell leukemia virus (HTLV-I) Rex trans-activator binds directly to the HTLV-I Rex and the type 1 human immunodeficiency virus Rev RNA response elements. *Proc. Natl. Acad. Sci. USA*, 88, 5704–5708.

Bolinger, C. and Boris-Lawrie, K. (2009). Mechanisms employed by retroviruses to exploit host factors for translational control of a complicated proteome. *Retrovirology*, 6, 8.

Booth, D.S., Cheng, Y., Frankel, A.D. (2014). The export receptor Crm1 forms a dimer to promote nuclear export of HIV RNA. *Elife*, 3, e04121.

Brasey, A., Lopez-Lastra, M., Ohlmann, T., Beerens, N., Berkhout, B., Darlix, J.-L., Sonenberg, N. (2003). The leader of human immunodeficiency virus type 1 genomic RNA harbors an internal ribosome entry segment that is active during the G2/M phase of the cell cycle. *J. Virol.*, 77, 3939–3949.

Braun, I.C., Rohrbach, E., Schmitt, C., Izaurralde, E. (1999). TAP binds to the constitutive transport element (CTE) through a novel RNA-binding motif that is sufficient to promote CTE-dependent RNA export from the nucleus. *EMBO J.*, 18, 1953–1965.

Bray, M., Prasad, S., Dubay, J.W., Hunter, E., Jeang, K.T., Rekosh, D., Hammarskjöld, M.L. (1994). A small element from the Mason-Pfizer monkey virus genome makes human immunodeficiency virus type 1 expression and replication Rev-independent. *Proc. Natl. Acad. Sci. USA*, 91, 1256–1260.

de Breyne, S. and Ohlmann, T. (2018). Focus on translation initiation of the HIV-1 mRNAs. *Int. J. Mol. Sci.*, 20, E101.

Buck, C.B., Shen, X., Egan, M.A., Pierson, T.C., Walker, C.M., Siliciano, R.F. (2001). The human immunodeficiency virus type 1 gag gene encodes an internal ribosome entry site. *J. Virol.*, 75, 181–191.

Burdick, R.C., Li, C., Munshi, M., Rawson, J.M.O., Nagashima, K., Hu, W.-S., Pathak, V.K. (2020). HIV-1 uncoats in the nucleus near sites of integration. *Proc. Natl. Acad. Sci. USA*, 117, 5486–5493.

Cabello-Villegas, J., Giles, K.E., Soto, A.M., Yu, P., Mougin, A., Beemon, K.L., Wang, Y.-X. (2004). Solution structure of the pseudo-5' splice site of a retroviral splicing suppressor. *RNA*, 10, 1388–1398.

Cáceres, C.J., Contreras, N., Angulo, J., Vera-Otarola, J., Pino-Ajenjo, C., Llorian, M., Ameur, M., Lisboa, F., Pino, K., Lowy, F. et al. (2016). Polypyrimidine tract-binding protein binds to the 5' untranslated region of the mouse mammary tumor virus mRNA and stimulates cap-independent translation initiation. *FEBS J.*, 283, 1880–1901.

Cáceres, C.J., Angulo, J., Lowy, F., Contreras, N., Walters, B., Olivares, E., Allouche, D., Merviel, A., Pino, K., Sargueil, B. et al. (2018). Non-canonical translation initiation of the spliced mRNA encoding the human T-cell leukemia virus type 1 basic leucine zipper protein. *Nucleic Acids Res.*, 46, 11030–11047.

Carpenter, S., Nadin-Davis, S.A., Wannemuehler, Y., Roth, J.A. (1993). Identification of transactivation-response sequences in the long terminal repeat of bovine immunodeficiency-like virus. *J. Virol.*, 67, 4399–4403.

Carvajal, F., Vallejos, M., Walters, B., Contreras, N., Hertz, M.I., Olivares, E., Cáceres, C.J., Pino, K., Letelier, A., Thompson, S.R. et al. (2016). Structural domains within the HIV-1 mRNA and the ribosomal protein S25 influence cap-independent translation initiation. *FEBS J.*, 283, 2508–2527.

Cavallari, I., Rende, F., Bona, M.K., Sztuba-Solinska, J., Silic-Benussi, M., Tognon, M., LeGrice, S.F.J., Franchini, G., D'Agostino, D.M., Ciminale, V. (2016). Expression of alternatively spliced human T-Cell leukemia virus type 1 mRNAs is influenced by mitosis and by a novel cis-acting regulatory sequence. *J. Virol.*, 90, 1486–1498.

Chameettachal, A., Vivet-Boudou, V., Pitchai, F.N.N., Pillai, V.N., Ali, L.M., Krishnan, A., Bernacchi, S., Mustafa, F., Marquet, R., Rizvi, T.A. (2021). A purine loop and the primer binding site are critical for the selective encapsidation of mouse mammary tumor virus genomic RNA by Pr77Gag. *Nucleic Acids Res.*, 49, 4668–4688.

Charnay, N., Ivanyi-Nagy, R., Soto-Rifo, R., Ohlmann, T., López-Lastra, M., Darlix, J.-L. (2009). Mechanism of HIV-1 Tat RNA translation and its activation by the Tat protein. *Retrovirology*, 6, 74.

Chavali, S.S., Bonn-Breach, R., Wedekind, J.E. (2019). Face-time with TAR: Portraits of an HIV-1 RNA with diverse modes of effector recognition relevant for drug discovery. *J. Biol. Chem.*, 294, 9326–9341.

Chen, H., Wilcox, G., Kertayadnya, G., Wood, C. (1999). Characterization of the Jembrana disease virus tat gene and the cis- and trans regulatory elements in its long terminal repeats. *J. Virol.*, 73, 658–666.

Chen, Y., Balakrishnan, M., Roques, B.P., Fay, P.J., Bambara, R.A. (2003). Mechanism of minus strand strong stop transfer in HIV-1 reverse transcription. *J. Biol. Chem.*, 278, 8006–8017.

Chen, J., Rahman, S.A., Nikolaitchik, O.A., Grunwald, D., Sardo, L., Burdick, R. C., Plisov, S., Liang, E., Tai, S., Pathak, V.K. et al. (2016a). HIV-1 RNA genome dimerizes on the plasma membrane in the presence of Gag protein. *Proc. Natl. Acad. Sci. USA*, 113, E201-208.

Chen, Y., Maskri, O., Chaminade, F., René, B., Benkaroun, J., Godet, J., Mély, Y., Mauffret, O., Fossé, P. (2016b). Structural insights into the HIV-1 minus-strand strong-stop DNA. *J. Biol. Chem.*, 291, 3468–3482.

Ciuffi, A. (2016). The benefits of integration. *Clin. Microbiol. Infect.*, 22, 324–332.

Clavel, F. and Orenstein, J.M. (1990). A mutant of human immunodeficiency virus with reduced RNA packaging and abnormal particle morphology. *J. Virol.*, 64, 5230–5234.

Clever, J.L. and Parslow, T.G. (1997). Mutant human immunodeficiency virus type 1 genomes with defects in RNA dimerization or encapsidation. *J. Virol.*, 71, 3407–3414.

Clever, J.L., Wong, M.L., Parslow, T.G. (1996). Requirements for kissing-loop-mediated dimerization of human immunodeficiency virus RNA. *J. Virol.*, 70, 5902–5908.

Cochrane, A.W., McNally, M.T., Mouland, A.J. (2006). The retrovirus RNA trafficking granule: From birth to maturity. *Retrovirology*, 3, 18.

Coffin, J.M. (1992). Retroviral DNA integration. *Dev. Biol. Stand.*, 76, 141–151.

Cosnefroy, O., Murray, P.J., Bishop, K.N. (2016). HIV-1 capsid uncoating initiates after the first strand transfer of reverse transcription. *Retrovirology*, 13, 58.

Cromer, D., Grimm, A.J., Schlub, T.E., Mak, J., Davenport, M.P. (2016). Estimating the in-vivo HIV template switching and recombination rate. *AIDS*, 30, 185–192.

D'Agostino, D.M., Cavallari, I., Romanelli, M.G., Ciminale, V. (2019). Post-transcriptional regulation of HTLV gene expression: Rex to the rescue. *Front. Microbiol.*, 10, 1958.

Darlix, J.L., Gabus, C., Nugeyre, M.T., Clavel, F., Barré-Sinoussi, F. (1990). Cis elements and trans-acting factors involved in the RNA dimerization of the human immunodeficiency virus HIV-1. *J. Mol. Biol.*, 216, 689–699.

Darlix, J.L., Vincent, A., Gabus, C., de Rocquigny, H., Roques, B. (1993). Trans-activation of the 5' to 3' viral DNA strand transfer by nucleocapsid protein during reverse transcription of HIV1 RNA. *C. R. Acad. Sci. III*, 316, 763–771.

Darlix, J.-L., Godet, J., Ivanyi-Nagy, R., Fossé, P., Mauffret, O., Mély, Y. (2011). Flexible nature and specific functions of the HIV-1 nucleocapsid protein. *J. Mol. Biol.*, 410, 565–581.

Das, A.T., Harwig, A., Vrolijk, M.M., Berkhout, B. (2007). The TAR hairpin of human immunodeficiency virus type 1 can be deleted when not required for Tat-mediated activation of transcription. *J. Virol.*, 81, 7742–7748.

Das, A.T., Vrolijk, M.M., Harwig, A., Berkhout, B. (2012). Opening of the TAR hairpin in the HIV-1 genome causes aberrant RNA dimerization and packaging. *Retrovirology*, 9, 59.

Daugherty, M.D., Booth, D.S., Jayaraman, B., Cheng, Y., Frankel, A.D. (2010a). HIV Rev response element (RRE) directs assembly of the Rev homooligomer into discrete asymmetric complexes. *Proc. Natl. Acad. Sci. USA*, 107, 12481–12486.

Daugherty, M.D., Liu, B., Frankel, A.D. (2010b). Structural basis for cooperative RNA binding and export complex assembly by HIV Rev. *Nat. Struct. Mol. Biol.*, 17, 1337–1342.

De Tapia, M., Metzler, V., Mougel, M., Ehresmann, B., Ehresmann, C. (1998). Dimerization of MoMuLV genomic RNA: Redefinition of the role of the palindromic stem-loop H1 (278-303) and new roles for stem-loops H2 (310-352) and H3 (355-374). *Biochemistry*, 37, 6077–6085.

Deffaud, C. and Darlix, J.L. (2000a). Characterization of an internal ribosomal entry segment in the 5' leader of murine leukemia virus env RNA. *J. Virol.*, 74, 846–850.

Deffaud, C. and Darlix, J.L. (2000b). Rous sarcoma virus translation revisited: Characterization of an internal ribosome entry segment in the 5' leader of the genomic RNA. *J. Virol.*, 74, 11581–11588.

Deforges, J., de Breyne, S., Ameur, M., Ulryck, N., Chamond, N., Saaidi, A., Ponty, Y., Ohlmann, T., Sargueil, B. (2017). Two ribosome recruitment sites direct multiple translation events within HIV1 Gag open reading frame. *Nucleic Acids Res.*, 45, 7382–7400.

Deng, H., Liu, R., Ellmeier, W., Choe, S., Unutmaz, D., Burkhart, M., Di Marzio, P., Marmon, S., Sutton, R.E., Hill, C.M. et al. (1996). Identification of a major co-receptor for primary isolates of HIV-1. *Nature*, 381, 661–666.

Derebail, S.S., Heath, M.J., DeStefano, J.J. (2003). Evidence for the differential effects of nucleocapsid protein on strand transfer in various regions of the HIV genome. *J. Biol. Chem.*, 278, 15702–15712.

Dirac, A.M., Huthoff, H., Kjems, J., Berkhout, B. (2001). The dimer initiation site hairpin mediates dimerization of the human immunodeficiency virus, type 2 RNA genome. *J. Biol. Chem.*, 276, 32345–32352.

Dismuke, D.J. and Aiken, C. (2006). Evidence for a functional link between uncoating of the human immunodeficiency virus type 1 core and nuclear import of the viral preintegration complex. *J. Virol.*, 80, 3712–3720.

Doria-Rose, N.A. and Vogt, V.M. (1998). In vivo selection of Rous sarcoma virus mutants with randomized sequences in the packaging signal. *J. Virol.*, 72, 8073–8082.

D'Orso, I. and Frankel, A.D. (2009). Tat acetylation modulates assembly of a viral-host RNA-protein transcription complex. *Proc. Natl. Acad. Sci. USA*, 106, 3101–3106.

Dostálková, A., Kaufman, F., Křížová, I., Kultová, A., Strohalmová, K., Hadravová, R., Ruml, T., Rumlová, M. (2018). Mutations in the basic region of the Mason-Pfizer monkey virus nucleocapsid protein affect reverse transcription, genomic RNA packaging, and the virus assembly site. *J. Virol.*, 92, e00106-18.

D'Souza, V. and Summers, M.F. (2004). Structural basis for packaging the dimeric genome of Moloney murine leukaemia virus. *Nature*, 431, 586–590.

D'Souza, V. and Summers, M.F. (2005). How retroviruses select their genomes. *Nat. Rev. Microbiol.*, 3, 643–655.

Dubois, N., Marquet, R., Paillart, J.-C., Bernacchi, S. (2018). Retroviral RNA dimerization: From structure to functions. *Front. Microbiol.*, 9, 527.

Ennifar, E., Walter, P., Ehresmann, B., Ehresmann, C., Dumas, P. (2001). Crystal structures of coaxially stacked kissing complexes of the HIV-1 RNA dimerization initiation site. *Nat. Struct. Biol.*, 8, 1064–1068.

Ernst, R.K., Bray, M., Rekosh, D., Hammarskjold, M.L. (1997). Secondary structure and mutational analysis of the Mason-Pfizer monkey virus RNA constitutive transport element. *RNA*, 3, 210–222.

Fabryova, H. and Strebel, K. (2019). Vpr and its cellular interaction partners: R we there yet? *Cells*, 8, 1310.

Fang, X., Wang, J., O'Carroll, I.P., Mitchell, M., Zuo, X., Wang, Y., Yu, P., Liu, Y., Rausch, J.W., Dyba, M.A. et al. (2013). An unusual topological structure of the HIV-1 Rev response element. *Cell*, 155, 594–605.

Feng, S. and Holland, E.C. (1988). HIV-1 tat trans-activation requires the loop sequence within tar. *Nature*, 334, 165–167.

Feng, Y., Broder, C.C., Kennedy, P.E., Berger, E.A. (1996). HIV-1 entry cofactor: Functional cDNA cloning of a seven-transmembrane, G protein-coupled receptor. *Science*, 272, 872–877.

Feng, Y.X., Campbell, S., Harvin, D., Ehresmann, B., Ehresmann, C., Rein, A. (1999). The human immunodeficiency virus type 1 Gag polyprotein has nucleic acid chaperone activity: Possible role in dimerization of genomic RNA and placement of tRNA on the primer binding site. *J. Virol.*, 73, 4251–4256.

Fenrick, R., Malim, M.H., Hauber, J., Le, S.Y., Maizel, J., Cullen, B.R. (1989). Functional analysis of the Tat trans activator of human immunodeficiency virus type 2. *J. Virol.*, 63, 5006–5012.

Fernandez, J., Machado, A.K., Lyonnais, S., Chamontin, C., Gärtner, K., Léger, T., Henriquet, C., Garcia, C., Portilho, D.M., Pugnière, M. et al. (2019). Transportin-1 binds to the HIV-1 capsid via a nuclear localization signal and triggers uncoating. *Nat. Microbiol.*, 4, 1840–1850.

Ferrer, M., Clerté, C., Chamontin, C., Basyuk, E., Lainé, S., Hottin, J., Bertrand, E., Margeat, E., Mougel, M. (2016). Imaging HIV-1 RNA dimerization in cells by multicolor super-resolution and fluctuation microscopies. *Nucleic Acids Res.*, 44, 7922–7934.

Flynn, J.A. and Telesnitsky, A. (2006). Two distinct Moloney murine leukemia virus RNAs produced from a single locus dimerize at random. *Virology*, 344, 391–400.

Fossé, P., Motté, N., Roumier, A., Gabus, C., Muriaux, D., Darlix, J.L., Paoletti, J. (1996). A short autocomplementary sequence plays an essential role in avian sarcoma-leukosis virus RNA dimerization. *Biochemistry*, 35, 16601–16609.

Fossé, P., Mougel, M., Keith, G., Westhof, E., Ehresmann, B., Ehresmann, C. (1998). Modified nucleotides of tRNAPro restrict interactions in the binary primer/template complex of M-MuLV. *J. Mol. Biol.*, 275, 731–746.

Fu, W. and Rein, A. (1993). Maturation of dimeric viral RNA of Moloney murine leukemia virus. *J. Virol.*, 67, 5443–5449.

Fu, W., Gorelick, R.J., Rein, A. (1994). Characterization of human immunodeficiency virus type 1 dimeric RNA from wild-type and protease-defective virions. *J. Virol.*, 68, 5013–5018.

Füzik, T., Píchalová, R., Schur, F.K.M., Strohalmová, K., Křížová, I., Hadravová, R., Rumlová, M., Briggs, J.A.G., Ulbrich, P., Ruml, T. (2016). Nucleic acid binding by Mason-Pfizer monkey virus CA promotes virus assembly and genome packaging. *J. Virol.*, 90, 4593–4603.

Galetto, R., Moumen, A., Giacomoni, V., Véron, M., Charneau, P., Negroni, M. (2004). The structure of HIV-1 genomic RNA in the gp120 gene determines a recombination hot spot in vivo. *J. Biol. Chem.*, 279, 36625–36632.

Gendron, K., Ferbeyre, G., Heveker, N., Brakier-Gingras, L. (2011). The activity of the HIV-1 IRES is stimulated by oxidative stress and controlled by a negative regulatory element. *Nucleic Acids Res.*, 39, 902–912.

Gherghe, C., Leonard, C.W., Gorelick, R.J., Weeks, K.M. (2010a). Secondary structure of the mature ex virio Moloney murine leukemia virus genomic RNA dimerization domain. *J. Virol.*, 84, 898–906.

Gherghe, C., Lombo, T., Leonard, C.W., Datta, S.A.K., Bess, J.W., Gorelick, R.J., Rein, A., Weeks, K.M. (2010b). Definition of a high-affinity Gag recognition structure mediating packaging of a retroviral RNA genome. *Proc. Natl. Acad. Sci. USA*, 107, 19248–19253.

Gilboa, E., Mitra, S.W., Goff, S., Baltimore, D. (1979). A detailed model of reverse transcription and tests of crucial aspects. *Cell*, 18, 93–100.

Girard, P.M., Bonnet-Mathonière, B., Muriaux, D., Paoletti, J. (1995). A short autocomplementary sequence in the 5' leader region is responsible for dimerization of MoMuLV genomic RNA. *Biochemistry*, 34, 9785–9794.

Girard, P.M., de Rocquigny, H., Roques, B.P., Paoletti, J. (1996). A model of PSI dimerization: Destabilization of the C278-G303 stem-loop by the nucleocapsid protein (NCp10) of MoMuLV. *Biochemistry*, 35, 8705–8714.

Gorelick, R.J., Henderson, L.E., Hanser, J.P., Rein, A. (1988). Point mutants of Moloney murine leukemia virus that fail to package viral RNA: Evidence for specific RNA recognition by a "zinc finger-like" protein sequence. *Proc. Natl. Acad. Sci. USA*, 85, 8420–8424.

Gorelick, R.J., Nigida, S.M., Bess, J.W., Arthur, L.O., Henderson, L.E., Rein, A. (1990). Noninfectious human immunodeficiency virus type 1 mutants deficient in genomic RNA. *J. Virol.*, 64, 3207–3211.

Gorelick, R.J., Chabot, D.J., Rein, A., Henderson, L.E., Arthur, L.O. (1993). The two zinc fingers in the human immunodeficiency virus type 1 nucleocapsid protein are not functionally equivalent. *J. Virol.*, 67, 4027–4036.

Greatorex, J.S., Laisse, V., Dockhelar, M.C., Lever, A.M. (1996). Sequences involved in the dimerisation of human T cell leukaemia virus type-1 RNA. *Nucleic Acids Res.*, 24, 2919–2923.

Grewe, B., Hoffmann, B., Ohs, I., Blissenbach, M., Brandt, S., Tippler, B., Grunwald, T., Uberla, K. (2012). Cytoplasmic utilization of human immunodeficiency virus type 1 genomic RNA is not dependent on a nuclear interaction with gag. *J. Virol.*, 86, 2990–3002.

Griffin, S.D., Allen, J.F., Lever, A.M. (2001). The major human immunodeficiency virus type 2 (HIV-2) packaging signal is present on all HIV-2 RNA species: Cotranslational RNA encapsidation and limitation of Gag protein confer specificity. *J. Virol.*, 75, 12058–12069.

Grohman, J.K., Gorelick, R.J., Kottegoda, S., Allbritton, N.L., Rein, A., Weeks, K.M. (2014). An immature retroviral RNA genome resembles a kinetically trapped intermediate state. *J. Virol.*, 88, 6061–6068.

Gröne, M., Hoffmann, E., Berchtold, S., Cullen, B.R., Grassmann, R. (1994). A single stem-loop structure within the HTLV-1 Rex response element is sufficient to mediate Rex activity in vivo. *Virology*, 204, 144–152.

Grüter, P., Tabernero, C., von Kobbe, C., Schmitt, C., Saavedra, C., Bachi, A., Wilm, M., Felber, B.K., Izaurralde, E. (1998). TAP, the human homolog of Mex67p, mediates CTE-dependent RNA export from the nucleus. *Mol. Cell.*, 1, 649–659.

Guo, J., Henderson, L.E., Bess, J., Kane, B., Levin, J.G. (1997). Human immunodeficiency virus type 1 nucleocapsid protein promotes efficient strand transfer and specific viral DNA synthesis by inhibiting TAR-dependent self-priming from minus-strand strong-stop DNA. *J. Virol.*, 71, 5178–5188.

Guo, F., Saadatmand, J., Niu, M., Kleiman, L. (2009). Roles of Gag and NCp7 in facilitating tRNA(Lys)(3) annealing to viral RNA in human immunodeficiency virus type 1. *J. Virol.*, 83, 8099–8107.

Hackett, P.B., Dalton, M.W., Johnson, D.P., Petersen, R.B. (1991). Phylogenetic and physical analysis of the 5' leader RNA sequences of avian retroviruses. *Nucleic Acids Res.*, 19, 6929–6934.

Hagan, N.A. and Fabris, D. (2007). Dissecting the protein–RNA and RNA–RNA interactions in the nucleocapsid-mediated dimerization and isomerization of HIV-1 stemloop 1. *J. Mol. Biol.*, 365, 396–410.

Hakata, Y., Umemoto, T., Matsushita, S., Shida, H. (1998). Involvement of human CRM1 (exportin 1) in the export and multimerization of the Rex protein of human T-cell leukemia virus type 1. *J. Virol.*, 72, 6602–6607.

Hanson, M.N., Balakrishnan, M., Roques, B.P., Bambara, R.A. (2005). Effects of donor and acceptor RNA structures on the mechanism of strand transfer by HIV-1 reverse transcriptase. *J. Mol. Biol.*, 353, 772–787.

Hendrix, J., Baumgärtel, V., Schrimpf, W., Ivanchenko, S., Digman, M.A., Gratton, E., Kräusslich, H.-G., Müller, B., Lamb, D.C. (2015). Live-cell observation of cytosolic HIV-1 assembly onset reveals RNA-interacting Gag oligomers. *J. Cell. Biol.*, 210, 629–646.

Heng, X., Kharytonchyk, S., Garcia, E.L., Lu, K., Divakaruni, S.S., LaCotti, C., Edme, K., Telesnitsky, A., Summers, M.F. (2012). Identification of a minimal region of the HIV-1 5'-leader required for RNA dimerization, NC binding, and packaging. *J. Mol. Biol.*, 417, 224–239.

Herbreteau, C.H., Weill, L., Décimo, D., Prévôt, D., Darlix, J.-L., Sargueil, B., Ohlmann, T. (2005). HIV-2 genomic RNA contains a novel type of IRES located downstream of its initiation codon. *Nat. Struct. Mol. Biol.*, 12, 1001–1007.

Hibbert, C.S., Mirro, J., Rein, A. (2004). mRNA molecules containing murine leukemia virus packaging signals are encapsidated as dimers. *J. Virol.*, 78, 10927–10938.

Höglund, S., Ohagen, A., Goncalves, J., Panganiban, A.T., Gabuzda, D. (1997). Ultrastructure of HIV-1 genomic RNA. *Virology*, 233, 271–279.

Hooker, C.W. and Harrich, D. (2003). The first strand transfer reaction of HIV-1 reverse transcription is more efficient in infected cells than in cell-free natural endogenous reverse transcription reactions. *J. Clin. Virol.*, 26, 229–238.

Houzet, L., Paillart, J.C., Smagulova, F., Maurel, S., Morichaud, Z., Marquet, R., Mougel, M. (2007). HIV controls the selective packaging of genomic, spliced viral and cellular RNAs into virions through different mechanisms. *Nucleic Acids Res.*, 35, 2695–2704.

Hudson, S.W., McNally, L.M., McNally, M.T. (2016). Evidence that a threshold of serine/arginine-rich (SR) proteins recruits CFIm to promote Rous sarcoma virus mRNA 3' end formation. *Virology*, 498, 181–191.

Hulme, A.E., Perez, O., Hope, T.J. (2011). Complementary assays reveal a relationship between HIV-1 uncoating and reverse transcription. *Proc. Natl. Acad. Sci. USA*, 108, 9975–9980.

Hussein, I.T.M., Ni, N., Galli, A., Chen, J., Moore, M.D., Hu, W.-S. (2010). Delineation of the preferences and requirements of the human immunodeficiency virus type 1 dimerization initiation signal by using an in vivo cell-based selection approach. *J. Virol.*, 84, 6866–6875.

Isel, C., Marquet, R., Keith, G., Ehresmann, C., Ehresmann, B. (1993). Modified nucleotides of tRNA(3Lys) modulate primer/template loop-loop interaction in the initiation complex of HIV-1 reverse transcription. *J. Biol. Chem.*, 268, 25269–25272.

Isel, C., Ehresmann, C., Keith, G., Ehresmann, B., Marquet, R. (1995). Initiation of reverse transcription of HIV-1: Secondary structure of the HIV-1 RNA/tRNA(3Lys) (template/primer). *J. Mol. Biol.*, 247, 236–250.

Isel, C., Ehresmann, C., Marquet, R. (2010). Initiation of HIV reverse transcription. *Viruses*, 2, 213–243.

Jaballah, S.A., Aktar, S.J., Ali, J., Phillip, P.S., Al Dhaheri, N.S., Jabeen, A., Rizvi, T.A. (2010). A G-C-rich palindromic structural motif and a stretch of single-stranded purines are required for optimal packaging of Mason-Pfizer monkey virus (MPMV) genomic RNA. *J. Mol. Biol.*, 401, 996–1014.

Jacks, T. and Varmus, H.E. (1985). Expression of the Rous sarcoma virus pol gene by ribosomal frameshifting. *Science*, 230, 1237–1242.

Jalalirad, M., Saadatmand, J., Laughrea, M. (2012). Dominant role of the 5' TAR bulge in dimerization of HIV-1 genomic RNA, but no evidence of TAR-TAR kissing during in vivo virus assembly. *Biochemistry*, 51, 3744–3758.

Jang, S.K., Kräusslich, H.G., Nicklin, M.J., Duke, G.M., Palmenberg, A.C., Wimmer, E. (1988). A segment of the 5' nontranslated region of encephalomyocarditis virus RNA directs internal entry of ribosomes during in vitro translation. *J. Virol.*, 62, 2636–2643.

Jin, D. and Musier-Forsyth, K. (2019). Role of host tRNAs and aminoacyl-tRNA synthetases in retroviral replication. *J. Biol. Chem.*, 294, 5352–5364.

Jossinet, F., Lodmell, J.S., Ehresmann, C., Ehresmann, B., Marquet, R. (2001). Identification of the in vitro HIV-2/SIV RNA dimerization site reveals striking differences with HIV-1. *J. Biol. Chem.*, 276, 5598–5604.

Jouvenet, N., Simon, S.M., Bieniasz, P.D. (2009). Imaging the interaction of HIV-1 genomes and Gag during assembly of individual viral particles. *Proc. Natl. Acad. Sci. USA*, 106, 19114–19119.

Jouvenet, N., Simon, S.M., Bieniasz, P.D. (2011). Visualizing HIV-1 assembly. *J. Mol. Biol.*, 410, 501–511.

Kalloush, R.M., Vivet-Boudou, V., Ali, L.M., Mustafa, F., Marquet, R., Rizvi, T.A. (2016). Packaging of Mason-Pfizer monkey virus (MPMV) genomic RNA depends upon conserved long-range interactions (LRIs) between U5 and gag sequences. *RNA*, 22, 905–919.

Kalloush, R.M., Vivet-Boudou, V., Ali, L.M., Pillai, V.N., Mustafa, F., Marquet, R., Rizvi, T.A. (2019). Stabilizing role of structural elements within the 5' untranslated region (UTR) and gag sequences in Mason-Pfizer monkey virus (MPMV) genomic RNA packaging. *RNA Biol.*, 16, 612–625.

Kaminska, M., Shalak, V., Francin, M., Mirande, M. (2007). Viral hijacking of mitochondrial lysyl-tRNA synthetase. *J. Virol.*, 81, 68–73.

Kanevsky, I., Chaminade, F., Ficheux, D., Moumen, A., Gorelick, R., Negroni, M., Darlix, J.-L., Fossé, P. (2005). Specific interactions between HIV-1 nucleocapsid protein and the TAR element. *J. Mol. Biol.*, 348, 1059–1077.

Kanevsky, I., Chaminade, F., Chen, Y., Godet, J., René, B., Darlix, J.-L., Mély, Y., Mauffret, O., Fossé, P. (2011). Structural determinants of TAR RNA-DNA annealing in the absence and presence of HIV-1 nucleocapsid protein. *Nucleic Acids Res.*, 39, 8148–8162.

Kang, Y. and Cullen, B.R. (1999). The human Tap protein is a nuclear mRNA export factor that contains novel RNA-binding and nucleocytoplasmic transport sequences. *Genes Dev.*, 13, 1126–1139.

Katahira, J. (2015). Nuclear export of messenger RNA. *Genes (Basel)*, 6, 163–184.

Katz, R.A. and Skalka, A.M. (1990). Control of retroviral RNA splicing through maintenance of suboptimal processing signals. *Mol. Cell. Biol.*, 10, 696–704.

Keane, S.C., Van, V., Frank, H.M., Sciandra, C.A., McCowin, S., Santos, J., Heng, X., Summers, M.F. (2016). NMR detection of intermolecular interaction sites in the dimeric 5'-leader of the HIV-1 genome. *Proc. Natl. Acad. Sci. USA*, 113, 13033–13038.

Keith, G. and Heyman, T. (1990). Heterogeneities in vertebrate tRNAs(Trp) avian retroviruses package only as a primer the tRNA(Trp) lacking modified m2G in position 7. *Nucleic Acids Res.*, 18, 703–710.

Kharytonchyk, S., Monti, S., Smaldino, P.J., Van, V., Bolden, N.C., Brown, J.D., Russo, E., Swanson, C., Shuey, A., Telesnitsky, A. et al. (2016). Transcriptional start site heterogeneity modulates the structure and function of the HIV-1 genome. *Proc. Natl. Acad. Sci. USA*, 113, 13378–13383.

Kharytonchyk, S., Brown, J.D., Stilger, K., Yasin, S., Iyer, A.S., Collins, J., Summers, M.F., Telesnitsky, A. (2018). Influence of gag and RRE Sequences on HIV-1 RNA Packaging Signal Structure and Function. *J. Mol. Biol.*, 430, 2066–2079.

Khoury, G., Mackenzie, C., Ayadi, L., Lewin, S.R., Branlant, C., Purcell, D.F.J. (2020). Tat IRES modulator of tat mRNA (TIM-TAM): A conserved RNA structure that controls Tat expression and acts as a switch for HIV productive and latent infection. *Nucleic Acids Res.*, 48, 2643–2660.

Kieken, F., Paquet, F., Brulé, F., Paoletti, J., Lancelot, G. (2006). A new NMR solution structure of the SL1 HIV-1Lai loop-loop dimer. *Nucleic Acids Res.*, 34, 343–352.

Klaver, B. and Berkhout, B. (1994). Premature strand transfer by the HIV-1 reverse transcriptase during strong-stop DNA synthesis. *Nucleic Acids Res.*, 22, 137–144.

Kleiman, L., Jones, C.P., Musier-Forsyth, K. (2010). Formation of the tRNALys packaging complex in HIV-1. *FEBS Lett.*, 584, 359–365.

Konvalinka, J., Kräusslich, H.-G., Müller, B. (2015). Retroviral proteases and their roles in virion maturation. *Virology*, 479–480, 403–417.

Kutluay, S.B. and Bieniasz, P.D. (2010). Analysis of the initiating events in HIV-1 particle assembly and genome packaging. *PloS Pathog.*, 6, e1001200.

Kuzembayeva, M., Dilley, K., Sardo, L., Hu, W.-S. (2014). Life of psi: How full-length HIV-1 RNAs become packaged genomes in the viral particles. *Virology*, 454–455, 362–370.

Lanchy, J.-M. and Lodmell, J.S. (2002). Alternate usage of two dimerization initiation sites in HIV-2 viral RNA in vitro. *J. Mol. Biol.*, 319, 637–648.

Lanchy, J.-M., Rentz, C.A., Ivanovitch, J.D., Lodmell, J.S. (2003). Elements located upstream and downstream of the major splice donor site influence the ability of HIV-2 leader RNA to dimerize in vitro. *Biochemistry*, 42, 2634–2642.

Laughrea, M. and Jetté, L. (1994). A 19-nucleotide sequence upstream of the 5' major splice donor is part of the dimerization domain of human immunodeficiency virus 1 genomic RNA. *Biochemistry*, 33, 13464–13474.

Laughrea, M. and Jetté, L. (1996). Kissing-loop model of HIV-1 genome dimerization: HIV-1 RNAs can assume alternative dimeric forms, and all sequences upstream or downstream of hairpin 248-271 are dispensable for dimer formation. *Biochemistry*, 35, 1589–1598.

Laughrea, M., Jetté, L., Mak, J., Kleiman, L., Liang, C., Wainberg, M.A. (1997). Mutations in the kissing-loop hairpin of human immunodeficiency virus type 1 reduce viral infectivity as well as genomic RNA packaging and dimerization. *J. Virol.*, 71, 3397–3406.

Le Blanc, I., Greatorex, J.S., Dokhélar, M.C., Lever, A.M. (2000). A 37 base sequence in the leader region of human T-cell leukaemia virus type I is a high affinity dimerization site but is not essential for virus replication. *J. Gen. Virol.*, 81, 105–108.

LeBlanc, J.J., Uddowla, S., Abraham, B., Clatterbuck, S., Beemon, K.L. (2007). Tap and Dbp5, but not Gag, are involved in DR-mediated nuclear export of unspliced Rous sarcoma virus RNA. *Virology*, 363, 376–386.

Lever, A.M., Gottlinger, H., Haseltine, W., Sodroski, J. (1989). Identification of a sequence required for efficient packaging of human immunodeficiency virus type 1 RNA into virions. *J. Virol.*, 63, 4085–4087.

Levin, J.G., Grimley, P.M., Ramseur, J.M., Berezesky, I.K. (1974). Deficiency of 60 to 70S RNA in murine leukemia virus particles assembled in cells treated with actinomycin D. *J. Virol.*, 14, 152–161.

Levin, J.G., Mitra, M., Mascarenhas, A., Musier-Forsyth, K. (2010). Role of HIV-1 nucleocapsid protein in HIV-1 reverse transcription. *RNA Biol.*, 7, 754–774.

L'Hernault, A., Greatorex, J.S., Crowther, R.A., Lever, A.M. (2007). Dimerisation of HIV-2 genomic RNA is linked to efficient RNA packaging, normal particle maturation and viral infectivity. *Retrovirology*, 4, 90.

L'Hernault, A., Weiss, E.U., Greatorex, J.S., Lever, A.M. (2012). HIV-2 genome dimerization is required for the correct processing of Gag: A second-site reversion in matrix can restore both processes in dimerization-impaired mutant viruses. *J. Virol.*, 86, 5867–5876.

Li, Y., Bor, Y.-C., Misawa, Y., Xue, Y., Rekosh, D., Hammarskjöld, M.-L. (2006). An intron with a constitutive transport element is retained in a Tap messenger RNA. *Nature*, 443, 234–237.

Lindemann, D., Steffen, I., Pöhlmann, S. (2013). Cellular entry of retroviruses. *Adv. Exp. Med. Biol.*, 790, 128–149.

Liu, S., Harada, B.T., Miller, J.T., Le Grice, S.F.J., Zhuang, X. (2010). Initiation complex dynamics direct the transitions between distinct phases of early HIV reverse transcription. *Nat. Struct. Mol. Biol.*, 17, 1453–1460.

Liu, Y., Nikolaitchik, O.A., Rahman, S.A., Chen, J., Pathak, V.K., Hu, W.-S. (2017). HIV-1 sequence necessary and sufficient to package non-viral RNAs into HIV-1 particles. *J. Mol. Biol.*, 429, 2542–2555.

Liu, S., Kaddis Maldonado, R., Rye-McCurdy, T., Binkley, C., Bah, A., Chen, E.C., Rice, B.L., Parent, L.J., Musier-Forsyth, K. (2020). Rous sarcoma virus genomic RNA dimerization capability in vitro is not a prerequisite for viral infectivity. *Viruses*, 12, E568.

Locker, N., Chamond, N., Sargueil, B. (2011). A conserved structure within the HIV gag open reading frame that controls translation initiation directly recruits the 40S subunit and eIF3. *Nucleic Acids Res.*, 39, 2367–2377.

López-Lastra, M., Gabus, C., Darlix, J.L. (1997). Characterization of an internal ribosomal entry segment within the 5' leader of avian reticuloendotheliosis virus type A RNA and development of novel MLV-REV-based retroviral vectors. *Hum. Gene Ther.*, 8, 1855–1865.

Lu, K., Heng, X., Garyu, L., Monti, S., Garcia, E.L., Kharytonchyk, S., Dorjsuren, B., Kulandaivel, G., Jones, S., Hiremath, A. et al. (2011). NMR detection of structures in the HIV-1 5'-leader RNA that regulate genome packaging. *Science*, 334, 242–245.

Lusvarghi, S., Sztuba-Solinska, J., Purzycka, K.J., Pauly, G.T., Rausch, J.W., Grice, S.F.J.L. (2013). The HIV-2 Rev-response element: Determining secondary structure and defining folding intermediates. *Nucleic Acids Res.*, 41, 6637–6649.

Ly, H. and Parslow, T.G. (2002). Bipartite signal for genomic RNA dimerization in Moloney murine leukemia virus. *J. Virol.*, 76, 3135–3144.

Mak, J. and Kleiman, L. (1997). Primer tRNAs for reverse transcription. *J. Virol.*, 71, 8087–8095.

Maldonado, R.J.K. and Parent, L.J. (2016). Orchestrating the selection and packaging of genomic RNA by retroviruses: An ensemble of viral and host factors. *Viruses*, 8, E257.

Maldonado, R.J.K., Rice, B., Chen, E.C., Tuffy, K.M., Chiari, E.F., Fahrbach, K. M., Hope, T.J., Parent, L.J. (2020). Visualizing association of the retroviral Gag protein with unspliced viral RNA in the nucleus. *Mbio*, 11, e00524-20.

Malim, M.H., Hauber, J., Le, S.Y., Maizel, J.V., Cullen, B.R. (1989). The HIV-1 rev trans-activator acts through a structured target sequence to activate nuclear export of unspliced viral mRNA. *Nature*, 338, 254–257.

Mamede, J.I., Cianci, G.C., Anderson, M.R., Hope, T.J. (2017). Early cytoplasmic uncoating is associated with infectivity of HIV-1. *Proc. Natl. Acad. Sci. USA*, 114, E7169–E7178.

Mangel, W.F., Delius, H., Duesberg, P.H. (1974). Structure and molecular weight of the 60-70S RNA and the 30-40S RNA of the Rous sarcoma virus. *Proc. Natl. Acad. Sci. USA*, 71, 4541–4545.

Mann, R., Mulligan, R.C., Baltimore, D. (1983). Construction of a retrovirus packaging mutant and its use to produce helper-free defective retrovirus. *Cell*, 33, 153–159.

Mansky, L.M. and Gajary, L.C. (2002). The primary nucleotide sequence of the bovine leukemia virus RNA packaging signal can influence efficient RNA packaging and virus replication. *Virology*, 301, 272–280.

Mansky, L.M. and Wisniewski, R.M. (1998). The bovine leukemia virus encapsidation signal is composed of RNA secondary structures. *J. Virol.*, 72, 3196–3204.

Mansky, L.M., Krueger, A.E., Temin, H.M. (1995). The bovine leukemia virus encapsidation signal is discontinuous and extends into the 5' end of the gag gene. *J. Virol.*, 69, 3282–3289.

Marquet, R., Baudin, F., Gabus, C., Darlix, J.L., Mougel, M., Ehresmann, C., Ehresmann, B. (1991). Dimerization of human immunodeficiency virus (type 1) RNA: Stimulation by cations and possible mechanism. *Nucleic Acids Res.*, 19, 2349–2357.

Marquet, R., Isel, C., Ehresmann, C., Ehresmann, B. (1995). tRNAs as primer of reverse transcriptases. *Biochimie*, 77, 113–124.

Masuda, T., Sato, Y., Huang, Y.-L., Koi, S., Takahata, T., Hasegawa, A., Kawai, G., Kannagi, M. (2015). Fate of HIV-1 cDNA intermediates during reverse transcription is dictated by transcription initiation site of virus genomic RNA. *Sci. Rep.*, 5, 17680.

Maurel, S. and Mougel, M. (2010). Murine leukemia virus RNA dimerization is coupled to transcription and splicing processes. *Retrovirology*, 7, 64.

Maurel, S., Houzet, L., Garcia, E.L., Telesnitsky, A., Mougel, M. (2007). Characterization of a natural heterodimer between MLV genomic RNA and the SD' retroelement generated by alternative splicing. *RNA*, 13, 2266–2276.

McBride, M.S. and Panganiban, A.T. (1997). Position dependence of functional hairpins important for human immunodeficiency virus type 1 RNA encapsidation in vivo. *J. Virol.*, 71, 2050–2058.

McCann, E.M. and Lever, A.M. (1997). Location of cis-acting signals important for RNA encapsidation in the leader sequence of human immunodeficiency virus type 2. *J. Virol.*, 71, 4133–4137.

McCauley, M.J., Rouzina, I., Manthei, K.A., Gorelick, R.J., Musier-Forsyth, K., Williams, M.C. (2015). Targeted binding of nucleocapsid protein transforms the folding landscape of HIV-1 TAR RNA. *Proc. Natl. Acad. Sci. USA*, 112, 13555–13560.

McNally, M.T. (2008). RNA processing control in avian retroviruses. *Front. Biosci.*, 13, 3869–3883.

McNally, L.M. and McNally, M.T. (1996). SR protein splicing factors interact with the Rous sarcoma virus negative regulator of splicing element. *J. Virol.*, 70, 1163–1172.

Méric, C. and Goff, S.P. (1989). Characterization of Moloney murine leukemia virus mutants with single-amino-acid substitutions in the Cys-His box of the nucleocapsid protein. *J. Virol.*, 63, 1558–1568.

Méric, C. and Spahr, P.F. (1986). Rous sarcoma virus nucleic acid-binding protein p12 is necessary for viral 70S RNA dimer formation and packaging. *J. Virol.*, 60, 450–459.

Méric, C., Gouilloud, E., Spahr, P.F. (1988). Mutations in Rous sarcoma virus nucleocapsid protein p12 (NC): Deletions of Cys-His boxes. *J. Virol.*, 62, 3328–3333.

Miller, S.B., Yildiz, F.Z., Lo, J.A., Wang, B., D'Souza, V.M. (2014). A structure-based mechanism for tRNA and retroviral RNA remodelling during primer annealing. *Nature*, 515, 591–595.

Miyazaki, Y., Irobalieva, R.N., Tolbert, B.S., Smalls-Mantey, A., Iyalla, K., Loeliger, K., D'Souza, V., Khant, H., Schmid, M.F., Garcia, E.L. et al. (2010). Structure of a conserved retroviral RNA packaging element by NMR spectroscopy and cryo-electron tomography. *J. Mol. Biol.*, 404, 751–772.

Monette, A., Ajamian, L., López-Lastra, M., Mouland, A.J. (2009). Human immunodeficiency virus type 1 (HIV-1) induces the cytoplasmic retention of heterogeneous nuclear ribonucleoprotein A1 by disrupting nuclear import: Implications for HIV-1 gene expression. *J. Biol. Chem.*, 284, 31350–31362.

Monette, A., Valiente-Echeverría, F., Rivero, M., Cohen, É.A., Lopez-Lastra, M., Mouland, A.J. (2013). Dual mechanisms of translation initiation of the full-length HIV-1 mRNA contribute to gag synthesis. *PloS One.*, 8, e68108.

Monie, T.P., Greatorex, J.S., Lever, A.M. (2001). Oligonucleotide mapping of the core genomic RNA dimer linkage in human T-cell leukaemia virus type-1. *Virus. Res.*, 78, 45–56.

Monie, T.P., Greatorex, J.S., Zacharias, M., Lever, A.M. (2004). The human T-cell lymphotropic virus type-I dimerization initiation site forms a hairpin loop, unlike previously characterized retroviral dimerization motifs. *Biochemistry*, 43, 6085–6090.

Moore, R., Dixon, M., Smith, R., Peters, G., Dickson, C. (1987). Complete nucleotide sequence of a milk-transmitted mouse mammary tumor virus: Two frameshift suppression events are required for translation of gag and pol. *J. Virol.*, 61, 480–490.

Moore, M.D., Fu, W., Nikolaitchik, O., Chen, J., Ptak, R.G., Hu, W.-S. (2007). Dimer initiation signal of human immunodeficiency virus type 1: Its role in partner selection during RNA copackaging and its effects on recombination. *J. Virol.*, 81, 4002–4011.

Moore, M.D., Nikolaitchik, O.A., Chen, J., Hammarskjöld, M.-L., Rekosh, D., Hu, W.-S. (2009). Probing the HIV-1 genomic RNA trafficking pathway and dimerization by genetic recombination and single virion analyses. *PloS Pathog.*, 5, e1000627.

Morcock, D.R., Katakam, S., Kane, B.P., Casas-Finet, J.R. (2002). Fluorescence and nucleic acid binding properties of bovine leukemia virus nucleocapsid protein. *Biophys. Chem.*, 97, 203–212.

Morris, S., Johnson, M., Stavnezer, E., Leis, J. (2002). Replication of avian sarcoma virus in vivo requires an interaction between the viral RNA and the TpsiC loop of the tRNA(Trp) primer. *J. Virol.*, 76, 7571–7577.

Mougel, M. and Barklis, E. (1997). A role for two hairpin structures as a core RNA encapsidation signal in murine leukemia virus virions. *J. Virol.*, 71, 8061–8065.

Mougel, M., Tounekti, N., Darlix, J.L., Paoletti, J., Ehresmann, B., Ehresmann, C. (1993). Conformational analysis of the 5' leader and the gag initiation site of Mo-MuLV RNA and allosteric transitions induced by dimerization. *Nucleic Acids Res.*, 21, 4677–4684.

Mougel, M., Zhang, Y., Barklis, E. (1996). Cis-active structural motifs involved in specific encapsidation of Moloney murine leukemia virus RNA. *J. Virol.*, 70, 5043–5050.

Mougel, M., Akkawi, C., Chamontin, C., Feuillard, J., Pessel-Vivares, L., Socol, M., Laine, S. (2020). NXF1 and CRM1 nuclear export pathways orchestrate nuclear export, translation and packaging of murine leukaemia retrovirus unspliced RNA. *RNA Biol.*, 17, 528–538.

Moumen, A., Polomack, L., Roques, B., Buc, H., Negroni, M. (2001). The HIV-1 repeated sequence R as a robust hot-spot for copy-choice recombination. *Nucleic Acids Res.*, 29, 3814–3821.

Moumen, A., Polomack, L., Unge, T., Véron, M., Buc, H., Negroni, M. (2003). Evidence for a mechanism of recombination during reverse transcription dependent on the structure of the acceptor RNA. *J. Biol. Chem.*, 278, 15973–15982.

Mujeeb, A., Ulyanov, N.B., Georgantis, S., Smirnov, I., Chung, J., Parslow, T.G., James, T.L. (2007). Nucleocapsid protein-mediated maturation of dimer initiation complex of full-length SL1 stemloop of HIV-1: Sequence effects and mechanism of RNA refolding. *Nucleic Acids Res.*, 35, 2026–2034.

Muriaux, D., Girard, P.M., Bonnet-Mathonière, B., Paoletti, J. (1995). Dimerization of HIV-1Lai RNA at low ionic strength. An autocomplementary sequence in the 5' leader region is evidenced by an antisense oligonucleotide. *J. Biol. Chem.*, 270, 8209–8216.

Muriaux, D., Fossé, P., Paoletti, J. (1996a). A kissing complex together with a stable dimer is involved in the HIV-1Lai RNA dimerization process in vitro. *Biochemistry*, 35, 5075–5082.

Muriaux, D., De Rocquigny, H., Roques, B.P., Paoletti, J. (1996b). NCp7 activates HIV-1Lai RNA dimerization by converting a transient loop–loop complex into a stable dimer. *J. Biol. Chem.*, 271, 33686–33692.

Murti, K.G., Bondurant, M., Tereba, A. (1981). Secondary structural features in the 70S RNAs of Moloney murine leukemia and Rous sarcoma viruses as observed by electron microscopy. *J. Virol.*, 37, 411–419.

Mustafa, F., Al Amri, D., Al Ali, F., Al Sari, N., Al Suwaidi, S., Jayanth, P., Philips, P.S., Rizvi, T.A. (2012). Sequences within both the 5' UTR and Gag are required for optimal in vivo packaging and propagation of mouse mammary tumor virus (MMTV) genomic RNA. *PloS One.*, 7, e47088.

Mustafa, F., Vivet-Boudou, V., Jabeen, A., Ali, L.M., Kalloush, R.M., Marquet, R., Rizvi, T.A. (2018). The bifurcated stem loop 4 (SL4) is crucial for efficient packaging of mouse mammary tumor virus (MMTV) genomic RNA. *RNA Biol.*, 15, 1047–1059.

Nakano, K. and Watanabe, T. (2016). HTLV-1 Rex tunes the cellular environment favorable for viral replication. *Viruses*, 8, 58.

Nam, S.H., Copeland, T.D., Hatanaka, M., Oroszlan, S. (1993). Characterization of ribosomal frameshifting for expression of pol gene products of human T-cell leukemia virus type I. *J. Virol.*, 67, 196–203.

Nash, M.A., Meyer, M.K., Decker, G.L., Arlinghaus, R.B. (1993). A subset of Pr65gag is nucleus associated in murine leukemia virus-infected cells. *J. Virol.*, 67, 1350–1356.

Negroni, M. and Buc, H. (2000). Copy-choice recombination by reverse transcriptases: Reshuffling of genetic markers mediated by RNA chaperones. *Proc. Natl. Acad. Sci. USA*, 97, 6385–6390.

Ni, N., Nikolaitchik, O.A., Dilley, K.A., Chen, J., Galli, A., Fu, W., Prasad, V.V.S.P., Ptak, R.G., Pathak, V.K., Hu, W.-S. (2011). Mechanisms of human immunodeficiency virus type 2 RNA packaging: Efficient trans packaging and selection of RNA copackaging partners. *J. Virol.*, 85, 7603–7612.

Nikolaitchik, O.A., Somoulay, X., Rawson, J.M.O., Yoo, J.A., Pathak, V.K., Hu, W.-S. (2020). Unpaired guanosines in the 5' untranslated region of HIV-1 RNA act synergistically to mediate genome packaging. *J. Virol.*, 94, e00439-20.

Ocwieja, K.E., Sherrill-Mix, S., Mukherjee, R., Custers-Allen, R., David, P., Brown, M., Wang, S., Link, D.R., Olson, J., Travers, K. et al. (2012). Dynamic regulation of HIV-1 mRNA populations analyzed by single-molecule enrichment and long-read sequencing. *Nucleic Acids Res.*, 40, 10345–10355.

Oertle, S. and Spahr, P.F. (1990). Role of the gag polyprotein precursor in packaging and maturation of Rous sarcoma virus genomic RNA. *J. Virol.*, 64, 5757–5763.

Ogert, R.A. and Beemon, K.L. (1998). Mutational analysis of the Rous sarcoma virus DR posttranscriptional control element. *J. Virol.*, 72, 3407–3411.

Ogert, R.A., Lee, L.H., Beemon, K.L. (1996). Avian retroviral RNA element promotes unspliced RNA accumulation in the cytoplasm. *J. Virol.*, 70, 3834–3843.

Ohi, Y. and Clever, J.L. (2000). Sequences in the 5' and 3' R elements of human immunodeficiency virus type 1 critical for efficient reverse transcription. *J. Virol.*, 74, 8324–8334.

Olivares, E., Landry, D.M., Cáceres, C.J., Pino, K., Rossi, F., Navarrete, C., Huidobro-Toro, J.P., Thompson, S.R., López-Lastra, M. (2014). The 5' untranslated region of the human T-cell lymphotropic virus type 1 mRNA enables cap-independent translation initiation. *J. Virol.*, 88, 5936–5955.

Olson, E.D., Cantara, W.A., Musier-Forsyth, K. (2015). New structure sheds light on selective HIV-1 genomic RNA packaging. *Viruses*, 7, 4826–4835.

Oroudjev, E.M., Kang, P.C., Kohlstaedt, L.A. (1999). An additional dimer linkage structure in Moloney murine leukemia virus RNA. *J. Mol. Biol.*, 291, 603–613.

Ott, M., Geyer, M., Zhou, Q. (2011). The control of HIV transcription: Keeping RNA polymerase II on track. *Cell Host Microbe*, 10, 426–435.

Paca, R.E., Ogert, R.A., Hibbert, C.S., Izaurralde, E., Beemon, K.L. (2000). Rous sarcoma virus DR posttranscriptional elements use a novel RNA export pathway. *J. Virol.*, 74, 9507–9514.

Paillart, J.C., Skripkin, E., Ehresmann, B., Ehresmann, C., Marquet, R. (1996a). A loop–loop "kissing" complex is the essential part of the dimer linkage of genomic HIV-1 RNA. *Proc. Natl. Acad. Sci. USA*, 93, 5572–5577.

Paillart, J.C., Berthoux, L., Ottmann, M., Darlix, J.L., Marquet, R., Ehresmann, B., Ehresmann, C. (1996b). A dual role of the putative RNA dimerization initiation site of human immunodeficiency virus type 1 in genomic RNA packaging and proviral DNA synthesis. *J. Virol.*, 70, 8348–8354.

Pallesen, J. (2011). Structure of the HIV-1 5' untranslated region dimer alone and in complex with gold nanocolloids: Support of a TAR-TAR-containing 5' dimer linkage site (DLS) and a 3' DIS-DIS-containing DLS. *Biochemistry*, 50, 6170–6177.

Panganiban, A.T. and Fiore, D. (1988). Ordered interstrand and intrastrand DNA transfer during reverse transcription. *Science*, 241, 1064–1069.

Pasquinelli, A.E., Ernst, R.K., Lund, E., Grimm, C., Zapp, M.L., Rekosh, D., Hammarskjöld, M.L., Dahlberg, J.E. (1997). The constitutive transport element (CTE) of Mason-Pfizer monkey virus (MPMV) accesses a cellular mRNA export pathway. *EMBO J.*, 16, 7500–7510.

Peliska, J.A. and Benkovic, S.J. (1992). Mechanism of DNA strand transfer reactions catalyzed by HIV-1 reverse transcriptase. *Science*, 258, 1112–1118.

Pelletier, J. and Sonenberg, N. (1988). Internal initiation of translation of eukaryotic mRNA directed by a sequence derived from poliovirus RNA. *Nature*, 334, 320–325.

Pessel-Vivares, L., Ferrer, M., Lainé, S., Mougel, M. (2014). MLV requires Tap/NXF1-dependent pathway to export its unspliced RNA to the cytoplasm and to express both spliced and unspliced RNAs. *Retrovirology*, 11, 21.

Pessel-Vivares, L., Houzet, L., Lainé, S., Mougel, M. (2015). Insights into the nuclear export of murine leukemia virus intron-containing RNA. *RNA Biol.*, 12, 942–949.

Pilkington, G.R., Purzycka, K.J., Bear, J., Le Grice, S.F.J., Felber, B.K. (2014). Gammaretrovirus mRNA expression is mediated by a novel, bipartite post-transcriptional regulatory element. *Nucleic Acids Res.*, 42, 11092–11106.

Pitchai, F.N.N., Chameettachal, A., Vivet-Boudou, V., Ali, L.M., Pillai, V.N., Krishnan, A., Bernacchi, S., Mustafa, F., Marquet, R., Rizvi, T.A. (2021). Identification of Pr78Gag binding sites on the Mason-Pfizer monkey virus genomic RNA packaging determinants. *J. Mol. Biol.*, 433, 166923.

Plank, T.-D.M., Whitehurst, J.T., Kieft, J.S. (2013). Cell type specificity and structural determinants of IRES activity from the 5' leaders of different HIV-1 transcripts. *Nucleic Acids Res.*, 41, 6698–6714.

Plank, T.-D.M., Whitehurst, J.T., Cencic, R., Pelletier, J., Kieft, J.S. (2014). Internal translation initiation from HIV-1 transcripts is conferred by a common RNA structure. *Translation (Austin)*, 2, e27694.

Poeschla, E., Gilbert, J., Li, X., Huang, S., Ho, A., Wong-Staal, F. (1998). Identification of a human immunodeficiency virus type 2 (HIV-2) encapsidation determinant and transduction of nondividing human cells by HIV-2-based lentivirus vectors. *J. Virol.*, 72, 6527–6536.

Polge, E., Darlix, J.L., Paoletti, J., Fossé, P. (2000). Characterization of loose and tight dimer forms of avian leukosis virus RNA. *J. Mol. Biol.*, 300, 41–56.

Pollom, E., Dang, K.K., Potter, E.L., Gorelick, R.J., Burch, C.L., Weeks, K.M., Swanstrom, R. (2013). Comparison of SIV and HIV-1 genomic RNA structures reveals impact of sequence evolution on conserved and non-conserved structural motifs. *PloS Pathog.*, 9, e1003294.

Pornillos, O. and Ganser-Pornillos, B.K. (2019). Maturation of retroviruses. *Curr. Opin. Virol.*, 36, 47–55.

Prats, A.C., Sarih, L., Gabus, C., Litvak, S., Keith, G., Darlix, J.L. (1988). Small finger protein of avian and murine retroviruses has nucleic acid annealing activity and positions the replication primer tRNA onto genomic RNA. *EMBO J.*, 7, 1777–1783.

Prats, A.C., Roy, C., Wang, P.A., Erard, M., Housset, V., Gabus, C., Paoletti, C., Darlix, J.L. (1990). Cis elements and trans-acting factors involved in dimer formation of murine leukemia virus RNA. *J. Virol.*, 64, 774–783.

Prchal, J., Kroupa, T., Ruml, T., Hrabal, R. (2013). Interaction of Mason-Pfizer monkey virus matrix protein with plasma membrane. *Front. Microbiol*, 4, 423.

Purzycka, K.J., Pachulska-Wieczorek, K., Adamiak, R.W. (2011). The in vitro loose dimer structure and rearrangements of the HIV-2 leader RNA. *Nucleic Acids Res.*, 39, 7234–7248.

Rausch, J.W. and Le Grice, S.F.J. (2015). HIV Rev assembly on the Rev response element (RRE): A structural perspective. *Viruses*, 7, 3053–3075.

René, B., Mauffret, O., Fossé, P. (2018). Retroviral nucleocapsid proteins and DNA strand transfers. *Biochim. Open*, 7, 10–25.

Rhim, H. and Rice, A.P. (1994). Functional significance of the dinucleotide bulge in stem-loop1 and stem-loop2 of HIV-2 TAR RNA. *Virology*, 202, 202–211.

Ricci, E.P., Herbreteau, C.H., Decimo, D., Schaupp, A., Datta, S.A.K., Rein, A., Darlix, J.-L., Ohlmann, T. (2008). In vitro expression of the HIV-2 genomic RNA is controlled by three distinct internal ribosome entry segments that are regulated by the HIV protease and the Gag polyprotein. *RNA*, 14, 1443–1455.

Rist, M.J. and Marino, J.P. (2002). Mechanism of nucleocapsid protein catalyzed structural isomerization of the dimerization initiation site of HIV-1. *Biochemistry*, 41, 14762–14770.

Rivas-Aravena, A., Ramdohr, P., Vallejos, M., Valiente-Echeverría, F., Dormoy-Raclet, V., Rodríguez, F., Pino, K., Holzmann, C., Huidobro-Toro, J.P., Gallouzi, I.-E. et al. (2009). The Elav-like protein HuR exerts translational control of viral internal ribosome entry sites. *Virology*, 392, 178–185.

Roda, R.H., Balakrishnan, M., Kim, J.K., Roques, B.P., Fay, P.J., Bambara, R.A. (2002). Strand transfer occurs in retroviruses by a pause-initiated two-step mechanism. *J. Biol. Chem.*, 277, 46900–46911.

Romani, B., Engelbrecht, S., Glashoff, R.H. (2010). Functions of Tat: The versatile protein of human immunodeficiency virus type 1. *J. Gen. Virol.*, 91, 1–12.

Roy, C., Tounekti, N., Mougel, M., Darlix, J.L., Paoletti, C., Ehresmann, C., Ehresmann, B., Paoletti, J. (1990). An analytical study of the dimerization of in vitro generated RNA of Moloney murine leukemia virus MoMuLV. *Nucleic Acids Res.*, 18, 7287–7292.

Sakuma, T., Tonne, J.M., Ikeda, Y. (2014a). Murine leukemia virus uses TREX components for efficient nuclear export of unspliced viral transcripts. *Viruses*, 6, 1135–1148.

Sakuma, T., Davila, J.I., Malcolm, J.A., Kocher, J.-P.A., Tonne, J.M., Ikeda, Y. (2014b). Murine leukemia virus uses NXF1 for nuclear export of spliced and unspliced viral transcripts. *J. Virol.*, 88, 4069–4082.

Schmidt, R.D., Mustafa, F., Lew, K.A., Browning, M.T., Rizvi, T.A. (2003). Sequences within both the 5' untranslated region and the gag gene are important for efficient encapsidation of Mason-Pfizer monkey virus RNA. *Virology*, 309, 166–178.

Schulze-Gahmen, U. and Hurley, J.H. (2018). Structural mechanism for HIV-1 TAR loop recognition by Tat and the super elongation complex. *Proc. Natl. Acad. Sci. USA*, 115, 12973–12978.

Schulze-Gahmen, U., Echeverria, I., Stjepanovic, G., Bai, Y., Lu, H., Schneidman-Duhovny, D., Doudna, J.A., Zhou, Q., Sali, A., Hurley, J.H. (2016). Insights into HIV-1 proviral transcription from integrative structure and dynamics of the Tat:AFF4:P-TEFb:TAR complex. *Elife*, 5, e15910.

Seif, E., Niu, M., Kleiman, L. (2015). In virio SHAPE analysis of tRNA(Lys3) annealing to HIV-1 genomic RNA in wild type and protease-deficient virus. *Retrovirology*, 12, 40.

Sertznig, H., Hillebrand, F., Erkelenz, S., Schaal, H., Widera, M. (2018). Behind the scenes of HIV-1 replication: Alternative splicing as the dependency factor on the quiet. *Virology*, 516, 176–188.

Shehu-Xhilaga, M., Crowe, S.M., Mak, J. (2001a). Maintenance of the Gag/Gag-Pol ratio is important for human immunodeficiency virus type 1 RNA dimerization and viral infectivity. *J. Virol.*, 75, 1834–1841.

Shehu-Xhilaga, M., Kraeusslich, H.G., Pettit, S., Swanstrom, R., Lee, J.Y., Marshall, J.A., Crowe, S.M., Mak, J. (2001b). Proteolytic processing of the p2/nucleocapsid cleavage site is critical for human immunodeficiency virus type 1 RNA dimer maturation. *J. Virol.*, 75, 9156–9164.

Shen, W., Gorelick, R.J., Bambara, R.A. (2011). HIV-1 nucleocapsid protein increases strand transfer recombination by promoting dimeric G-quartet formation. *J. Biol. Chem.*, 286, 29838–29847.

Sherpa, C., Rausch, J.W., Le Grice, S.F.J., Hammarskjold, M.-L., Rekosh, D. (2015). The HIV-1 Rev response element (RRE) adopts alternative conformations that promote different rates of virus replication. *Nucleic Acids Res.*, 43, 4676–4686.

Shida, H. (2012). Role of nucleocytoplasmic RNA transport during the life cycle of retroviruses. *Front. Microbiol.*, 3, 179.

Simon-Loriere, E., Rossolillo, P., Negroni, M. (2011). RNA structures, genomic organization and selection of recombinant HIV. *RNA Biol.*, 8, 280–286.

Simpson, S.B., Zhang, L., Craven, R.C., Stoltzfus, C.M. (1997). Rous sarcoma virus direct repeat cis elements exert effects at several points in the virus life cycle. *J. Virol.*, 71, 9150–9156.

Skripkin, E., Paillart, J.C., Marquet, R., Ehresmann, B., Ehresmann, C. (1994). Identification of the primary site of the human immunodeficiency virus type 1 RNA dimerization in vitro. *Proc. Natl. Acad. Sci. USA*, 91, 4945–4949.

Sleiman, D., Goldschmidt, V., Barraud, P., Marquet, R., Paillart, J.-C., Tisné, C. (2012). Initiation of HIV-1 reverse transcription and functional role of nucleocapsid-mediated tRNA/viral genome interactions. *Virus Res.*, 169, 324–339.

Smagulova, F., Maurel, S., Morichaud, Z., Devaux, C., Mougel, M., Houzet, L. (2005). The highly structured encapsidation signal of MuLV RNA is involved in the nuclear export of its unspliced RNA. *J. Mol. Biol.*, 354, 1118–1128.

Sommerfelt, M.A. (1999). Retrovirus receptors. *J. Gen. Virol.*, 80(12), 3049–3064.

Steckbeck, J.D., Kuhlmann, A.-S., Montelaro, R.C. (2014). Structural and functional comparisons of retroviral envelope protein C-terminal domains: Still much to learn. *Viruses*, 6, 284–300.

Stewart, L., Schatz, G., Vogt, V.M. (1990). Properties of avian retrovirus particles defective in viral protease. *J. Virol.*, 64, 5076–5092.

Stoltzfus, C.M. and Snyder, P.N. (1975). Structure of B77 sarcoma virus RNA: Stabilization of RNA after packaging. *J. Virol.*, 16, 1161–1170.

Tanese, N., Telesnitsky, A., Goff, S.P. (1991). Abortive reverse transcription by mutants of Moloney murine leukemia virus deficient in the reverse transcriptase-associated RNase H function. *J. Virol.*, 65, 4387–4397.

Temin, H.M. and Mizutani, S. (1970). RNA-dependent DNA polymerase in virions of Rous sarcoma virus. *Nature*, 226, 1211–1213.

Teplova, M., Wohlbold, L., Khin, N.W., Izaurralde, E., Patel, D.J. (2011). Structure–function studies of nucleocytoplasmic transport of retroviral genomic RNA by mRNA export factor TAP. *Nat. Struct. Mol. Biol.*, 18, 990–998.

Thomas, J.A. and Gorelick, R.J. (2008). Nucleocapsid protein function in early infection processes. *Virus Res.*, 134, 39–63.

Tisné, C., Roques, B.P., Dardel, F. (2004). The annealing mechanism of HIV-1 reverse transcription primer onto the viral genome. *J. Biol. Chem.*, 279, 3588–3595.

Tomezsko, P.J., Corbin, V.D.A., Gupta, P., Swaminathan, H., Glasgow, M., Persad, S., Edwards, M.D., Mcintosh, L., Papenfuss, A.T., Emery, A. et al. (2020). Determination of RNA structural diversity and its role in HIV-1 RNA splicing. *Nature*, 582, 438–442.

Tounekti, N., Mougel, M., Roy, C., Marquet, R., Darlix, J.L., Paoletti, J., Ehresmann, B., Ehresmann, C. (1992). Effect of dimerization on the conformation of the encapsidation Psi domain of Moloney murine leukemia virus RNA. *J. Mol. Biol.*, 223, 205–220.

Toyoshima, H., Itoh, M., Inoue, J., Seiki, M., Takaku, F., Yoshida, M. (1990). Secondary structure of the human T-cell leukemia virus type 1 rex-responsive element is essential for rex regulation of RNA processing and transport of unspliced RNAs. *J. Virol.*, 64, 2825–2832.

Tran, T., Liu, Y., Marchant, J., Monti, S., Seu, M., Zaki, J., Yang, A.L., Bohn, J., Ramakrishnan, V., Singh, R. et al. (2015). Conserved determinants of lentiviral genome dimerization. *Retrovirology*, 12, 83.

Trono, D. (1992). Partial reverse transcripts in virions from human immunodeficiency and murine leukemia viruses. *J. Virol.*, 66, 4893–4900.

Trubetskoy, A.M., Okenquist, S.A., Lenz, J. (1999). R region sequences in the long terminal repeat of a murine retrovirus specifically increase expression of unspliced RNAs. *J. Virol.*, 73, 3477–3483.

Tuffy, K.M., Maldonado, R.J.K., Chang, J., Rosenfeld, P., Cochrane, A., Parent, L.J. (2020). HIV-1 Gag forms ribonucleoprotein complexes with unspliced viral RNA at transcription sites. *Viruses*, 12, E1281.

Ukah, O.B., Puray-Chavez, M., Tedbury, P.R., Herschhorn, A., Sodroski, J.G., Sarafianos, S.G. (2018). Visualization of HIV-1 RNA transcription from integrated HIV-1 DNA in reactivated latently infected cells. *Viruses*, 10, E534.

Ulyanov, N.B., Mujeeb, A., Du, Z., Tonelli, M., Parslow, T.G., James, T.L. (2006). NMR structure of the full-length linear dimer of stem-loop-1 RNA in the HIV-1 dimer initiation site. *J. Biol. Chem.*, 281, 16168–16177.

Umunnakwe, C.N., Duchon, A., Nikolaitchik, O.A., Rahman, S.A., Liu, Y., Chen, J., Tai, S., Pathak, V.K., Hu, W.-S. (2021). Specific guanosines in the HIV-2 leader RNA are essential for efficient viral genome packaging. *J. Mol. Biol.*, 433, 166718.

Vagner, S., Waysbort, A., Marenda, M., Gensac, M.C., Amalric, F., Prats, A.C. (1995). Alternative translation initiation of the Moloney murine leukemia virus mRNA controlled by internal ribosome entry involving the p57/PTB splicing factor. *J. Biol. Chem.*, 270, 20376–20383.

Vallejos, M., Ramdohr, P., Valiente-Echeverría, F., Tapia, K., Rodriguez, F.E., Lowy, F., Huidobro-Toro, J.P., Dangerfield, J.A., López-Lastra, M. (2010). The 5'-untranslated region of the mouse mammary tumor virus mRNA exhibits cap-independent translation initiation. *Nucleic Acids Res.*, 38, 618–632.

Vallejos, M., Deforges, J., Plank, T.-D.M., Letelier, A., Ramdohr, P., Abraham, C.G., Valiente-Echeverría, F., Kieft, J.S., Sargueil, B., López-Lastra, M. (2011). Activity of the human immunodeficiency virus type 1 cell cycle-dependent internal ribosomal entry site is modulated by IRES trans-acting factors. *Nucleic Acids Res.*, 39, 6186–6200.

Viphakone, N., Hautbergue, G.M., Walsh, M., Chang, C.-T., Holland, A., Folco, E.G., Reed, R., Wilson, S.A. (2012). TREX exposes the RNA-binding domain of Nxf1 to enable mRNA export. *Nat. Commun.*, 3, 1006.

Vo, M.-N., Barany, G., Rouzina, I., Musier-Forsyth, K. (2009). HIV-1 nucleocapsid protein switches the pathway of transactivation response element RNA/DNA annealing from loop-loop “kissing” to “zipper”. *J. Mol. Biol.*, 386, 789–801.

Vogt, V.M. (1997). Retroviral virions and genomes. In *Retroviruses*, Coffin, J.M., Hughes, S.H., Varmus, H.E. (eds). Cold Spring Harbor Laboratory Press, New York.

Volkova, N.A., Fomina, E.G., Smolnikova, V.V., Zinovieva, N.A., Fomin, I.K. (2014). The U3 region of Moloney murine leukemia virus contains position-independent cis-acting sequences involved in the nuclear export of full-length viral transcripts. *J. Biol. Chem.*, 289, 20158–20169.

Wang, H., Norris, K.M., Mansky, L.M. (2003). Involvement of the matrix and nucleocapsid domains of the bovine leukemia virus Gag polyprotein precursor in viral RNA packaging. *J. Virol.*, 77, 9431–9438.

Wang, B., Rekosh, D., Hammarskjold, M.-L. (2015). Evolutionary conservation of a molecular machinery for export and expression of mRNAs with retained introns. *RNA*, 21, 426–437.

Watts, J.M., Dang, K.K., Gorelick, R.J., Leonard, C.W., Bess, J.W., Swanstrom, R., Burch, C.L., Weeks, K.M. (2009). Architecture and secondary structure of an entire HIV-1 RNA genome. *Nature*, 460, 711–716.

Weill, L., James, L., Ulryck, N., Chamond, N., Herbreteau, C.H., Ohlmann, T., Sargueil, B. (2010). A new type of IRES within gag coding region recruits three initiation complexes on HIV-2 genomic RNA. *Nucleic Acids Res.*, 38, 1367–1381.

Werner, S., Vogel-Bachmayr, K., Hollinderbäumer, B., Wöhrl, B.M. (2001). Requirements for minus-strand transfer catalyzed by Rous sarcoma virus reverse transcriptase. *J. Virol.*, 75, 10132–10138.

Wilkinson, K.A., Gorelick, R.J., Vasa, S.M., Guex, N., Rein, A., Mathews, D.H., Giddings, M.C., Weeks, K.M. (2008). High-throughput SHAPE analysis reveals structures in HIV-1 genomic RNA strongly conserved across distinct biological states. *PloS Biol.*, 6, e96.

Wilson, W., Braddock, M., Adams, S.E., Rathjen, P.D., Kingsman, S.M., Kingsman, A.J. (1988). HIV expression strategies: Ribosomal frameshifting is directed by a short sequence in both mammalian and yeast systems. *Cell*, 55, 1159–1169.

Wu, W., Hatterschide, J., Syu, Y.-C., Cantara, W.A., Blower, R.J., Hanson, H.M., Mansky, L.M., Musier-Forsyth, K. (2018). Human T-cell leukemia virus type 1 Gag domains have distinct RNA-binding specificities with implications for RNA packaging and dimerization. *J. Biol. Chem.*, 293, 16261–16276.

Xie, L., Kesic, M., Yamamoto, B., Li, M., Younis, I., Lairmore, M.D., Green, P.L. (2009). Human T-cell leukemia virus type 2 Rex carboxy terminus is an inhibitory/stability domain that regulates Rex functional activity and viral replication. *J. Virol.*, 83, 5232–5243.

Xing, L., Liang, C., Kleiman, L. (2011). Coordinate roles of Gag and RNA helicase A in promoting the annealing of formula to HIV-1 RNA. *J. Virol.*, 85, 1847–1860.

Xing, L., Niu, M., Kleiman, L. (2012). In vitro and in vivo analysis of the interaction between RNA helicase A and HIV-1 RNA. *J. Virol.*, 86, 13272–13280.

Yang, J. and Cullen, B.R. (1999). Structural and functional analysis of the avian leukemia virus constitutive transport element. *RNA*, 5, 1645–1655.

Yoshinaka, Y., Katoh, I., Copeland, T.D., Oroszlan, S. (1985a). Murine leukemia virus protease is encoded by the gag-pol gene and is synthesized through suppression of an amber termination codon. *Proc. Natl. Acad. Sci. USA*, 82, 1618–1622.

Yoshinaka, Y., Katoh, I., Copeland, T.D., Oroszlan, S. (1985b). Translational readthrough of an amber termination codon during synthesis of feline leukemia virus protease. *J. Virol.*, 55, 870–873.

Yu, H., Jetzt, A.E., Ron, Y., Preston, B.D., Dougherty, J.P. (1998). The nature of human immunodeficiency virus type 1 strand transfers. *J. Biol. Chem.*, 273, 28384–28391.

Zhang, Y. and Barklis, E. (1995). Nucleocapsid protein effects on the specificity of retrovirus RNA encapsidation. *J. Virol.*, 69, 5716–5722.

Zhang, L. and Stoltzfus, C.M. (1995). A suboptimal src 3' splice site is necessary for efficient replication of Rous sarcoma virus. *Virology*, 206, 1099–1107.

Zhang, H., Zhang, Y., Spicer, T.P., Abbott, L.Z., Abbott, M., Poiesz, B.J. (1993). Reverse transcription takes place within extracellular HIV-1 virions: Potential biological significance. *AIDS Res. Hum. Retroviruses*, 9, 1287–1296.

Zhang, W., Cao, S., Martin, J.L., Mueller, J.D., Mansky, L.M. (2015). Morphology and ultrastructure of retrovirus particles. *AIMS Biophys.*, 2, 343–369.

Zhou, J., Bean, R.L., Vogt, V.M., Summers, M. (2007). Solution structure of the Rous sarcoma virus nucleocapsid protein: muPsi RNA packaging signal complex. *J. Mol. Biol.*, 365, 453–467.

Zolotukhin, A.S., Harford, J.B., Felber, B.K. (1994). Rev of human immunodeficiency virus and Rex of the human T-cell leukemia virus type I can counteract an mRNA downregulatory element of the transferrin receptor mRNA. *Nucleic Acids Res.*, 22, 4725–4732.

Index

A, B, C

D, E

G, H

I, L

N, O

P, R

S, T

U, V, X

Other titles from

iSTE

in

Biology and Biomedical Engineering

2022

DAMBRICOURT MALASSÉ Anne
Embryogeny and Phylogeny of the Human Posture 2: A New Glance at the Future of our Species (Comparative Anatomy and Posture of Animal and Human Set – Volume 4)

2021

DAMBRICOURT MALASSÉ Anne
Embryogeny and Phylogeny of the Human Posture 1: A New Glance at the Future of our Species (Comparative Anatomy and Posture of Animal and Human Set – Volume 3)

GRANDCOLAS Philippe, MAUREL Marie-Christine
Systematics and the Exploration of Life

HADJOUIS Djillali
The Skull of Quadruped and Bipedal Vertebrates: Variations, Abnormalities and Joint Pathologies (Comparative Anatomy and Posture of Animal and Human Set – Volume 2)

HULLÉ Maurice, VERNON Philippe
The Terrestrial Macroinvertebrates of the Sub-Antarctic Îles Kerguelen and Île de la Possession

2020

CAZEAU Cyrille
Foot Surgery Viewed Through the Prism of Comparative Anatomy: From Normal to Useful
(Comparative Anatomy and Posture of Animal and Human Set – Volume 1)

DUJON Bernard, PELLETIER Georges
Trajectories of Genetics

2019

BRAND Gérard
Discovering Odors

BUIS Roger
Biology and Mathematics: History and Challenges

2016

CLERC Maureen, BOUGRAIN Laurent, LOTTE Fabien
Brain-Computer Interfaces 1: Foundations and Methods
Brain-Computer Interfaces 2: Technology and Applications

FURGER Christophe
Live Cell Assays: From Research to Health and Regulatory Applications

2015

CLARYSSE Patrick, FRIBOULET Denis
Multi-modality Cardiac Imaging: Processing and Analysis

2014

CHEZE Laurence
Kinematic Analysis of Human Movement

CLARYSSE Patrick, FRIBOULET Denis
Multi-modality Cardiac Imaging: Processing and Analysis

DAO Tien Tuan, HO BA THO Marie-Christine
Biomechanics of the Musculoskeletal: Modeling of Data Uncertainty and Knowledge

FANET Hervé
Medical Imaging Based on Magnetic Fields and Ultrasounds

FARINAS DEL CERRO Luis, INOUE Katsumi
Logical Modeling of Biological Systems

MIGONNEY Véronique
Biomaterials

TEBBANI Sihem, LOPES Filipa, FILALI Rayen, DUMUR Didier, PAREAU Dominique
CO_2 Biofixation by Microalgae: Modeling, Estimation and Control